AF389457

Advances in Antimicrobial Coatings

Advances in Antimicrobial Coatings

Editor

Sami Rtimi

MDPI • Basel • Beijing • Wuhan • Barcelona • Belgrade • Manchester • Tokyo • Cluj • Tianjin

Editor
Sami Rtimi
Swiss Federal Institute of Technology Lausanne (EPFL)
Switzerland
rtimi.sami@gmail.com

Editorial Office
MDPI
St. Alban-Anlage 66
4052 Basel, Switzerland

This is a reprint of articles from the Special Issue published online in the open access journal *Coatings* (ISSN 2079-6412) (available at: https://www.mdpi.com/journal/coatings/special_issues/Adv_Anti_Coat).

For citation purposes, cite each article independently as indicated on the article page online and as indicated below:

LastName, A.A.; LastName, B.B.; LastName, C.C. Article Title. *Journal Name* **Year**, *Volume Number*, Page Range.

ISBN 978-3-0365-1006-4 (Hbk)
ISBN 978-3-0365-1007-1 (PDF)

Contents

About the Editor

Sami Rtimi (Ph.D., Ph.D.) was awarded a Ph.D. in Chemistry and Chemical Engineering from the Swiss Federal Institute of Technology (EPFL) and a Doctorate in Biological Sciences from the University of Carthage. He is specialist in green solutions for indoor environmental remediation. With an h-index of 30, he has published more than 130 articles in peer-reviewed journals, patents, and several book chapters and presented numerous communications at international meetings. According to Scopus, he is the second most active scientist worldwide working on "Disinfection; Photocatalysis; Microbial inactivation". He has also served as a reviewer of international grants (Switzerland, Canada, Greece, Spain, Poland, Chile, etc.). He has helped in the organization of a number of scientific meetings/conferences/workshops and chaired several sessions. He has been active in several NGOs to promote sustainability solutions (water and health-related issues) in Least Developed Countries (LDCs).

References

Klein, S.E.; Alzagameem, A.; Rumpf, J.; Korte, I.; Kreyenschmidt, J.; Schulze, M. Antimicrobial Activity of Lignin-Derived Polyurethane Coatings Prepared from Unmodified and Demethylated Lignins. *Coatings* **2019**, *9*, 494. [CrossRef]

Sanbhal, N.; Li, Y.; Khatri, A.; Peerzada, M.; Wang, L. Chitosan Cross-Linked Bio-based Antimicrobial Polypropylene Meshes for Hernia Repair Loaded with Levofloxacin HCl via Cold Oxygen Plasma. *Coatings* **2019**, *9*, 168. [CrossRef]

Saitaer, X.; Sanbhal, N.; Qiao, Y.; Li, Y.; Gao, J.; Brochu, G.; Guidoin, R.; Khatri, A.; Wang, L. Polydopamine-Inspired Surface Modification of Polypropylene Hernia Mesh Devices via Cold Oxygen Plasma: Antibacterial and Drug Release Properties. *Coatings* **2019**, *9*, 164. [CrossRef]

Kojima, R.; Kobayashi, S.; Matsumura, K.; Satuito, C.G.P.; Seki, Y.; Ando, H.; Katsuyama, I. Designing a Laboratory Bioassay for Evaluating the Efficacy of Antifouling Paints on Amphibalanus amphitrite Using a Flow-Through System. *Coatings* **2019**, *9*, 112. [CrossRef]

Nokkrut, B.O.; Pisuttipiched, S.; Khantayanuwong, S.; Puangsin, B. Silver Nanoparticle-Based Paper Packaging to Combat Black Anther Disease in Orchid Flowers. *Coatings* **2019**, *9*, 40. [CrossRef]

Druvari, D.; Koromilas, N.D.; Bekiari, V.; Bokias, G.; Kallitsis, J.K. Polymeric Antimicrobial Coatings Based on Quaternary Ammonium Compounds. *Coatings* **2018**, *8*, 8. [CrossRef]

Santonicola, S.; García Ibarra, V.; Sendón, R.; Mercogliano, R.; Rodríguez-Bernaldo de Quirós, A. Antimicrobial Films Based on Chitosan and Methylcellulose Containing Natamycin for Active Packaging Applications. *Coatings* **2017**, *7*, 177. [CrossRef]

Mathew, E.; Domínguez-Robles, J.; Larrañeta, E.; Lamprou, D.A. Fused Deposition Modelling as a Potential Tool for Antimicrobial Dialysis Catheters Manufacturing: New Trends vs. Conventional Approaches. *Coatings* **2019**, *9*, 515. [CrossRef]

Socaciu, M.-I.; Semeniuc, C.A.; Vodnar, D.C. Edible Films and Coatings for Fresh Fish Packaging: Focus on Quality Changes and Shelf-life Extension. *Coatings* **2018**, *8*, 366. [CrossRef]

Alhussein, A.; Achache, S.; Deturche, R.; Rtimi, S. Architectured Cu–TNTZ Bilayered Coatings Showing Bacterial Inactivation under Indoor Light and Controllable Copper Release: Effect of the Microstructure on Copper Diffusion. *Coatings* **2020**, *10*, 574. [CrossRef]

Vasilev, K. Nanoengineered Antibacterial Coatings and Materials: A Perspective. *Coatings* **2019**, *9*, 654. [CrossRef]

Article

Antimicrobial Activity of Lignin-Derived Polyurethane Coatings Prepared from Unmodified and Demethylated Lignins

Stephanie Elisabeth Klein [1,2], Abla Alzagameem [1,3], Jessica Rumpf [1,4]Imke Korte [4], Judith Kreyenschmidt [4] and Margit Schulze [1,*]

[1] Department of Natural Sciences, Bonn-Rhein-Sieg University of Applied Sciences, von-Liebig-Str. 20, D-53359 Rheinbach, Germany
[2] Department of Macromolecular Chemistry, Technical University of Darmstadt, Alarich-Weiß-Straße 4, D-64287 Darmstadt, Germany
[3] Faculty of Environment and Natural Sciences, Brandenburg University of Technology BTU Cottbus-Senftenberg, Platz der Deutschen Einheit 1, D-03046 Cottbus, Germany
[4] Faculty of Agriculture, Rheinische Friedrich Wilhelms-University Bonn, Katzenburgweg 7-9, D-53115 Bonn, Germany
* Correspondence: margit.schulze@h-brs.de; Tel.: +49-2241-865-566; Fax: +49-2241-865-8566

Received: 24 May 2019; Accepted: 26 July 2019; Published: 5 August 2019

Abstract: Due to global ecological and economic challenges that have been correlated to the transition from fossil-based to renewable resources, fundamental studies are being performed worldwide to replace fossil fuel raw materials in plastic production. One aspect of current research is the development of lignin-derived polyols to substitute expensive fossil-based polyol components for polyurethane and polyester production. This article describes the synthesis of bioactive lignin-based polyurethane coatings using unmodified and demethylated Kraft lignins. Demethylation was performed to enhance the reaction selectivity toward polyurethane formation. The antimicrobial activity was tested according to a slightly modified standard test (JIS Z 2801:2010). Besides effects caused by the lignins themselves, triphenylmethane derivatives (brilliant green and crystal violet) were used as additional antimicrobial substances. Results showed increased antimicrobial capacity against *Staphylococcus aureus*. Furthermore, the coating color could be varied from dark brown to green and blue, respectively.

Keywords: antimicrobial activity; brilliant green; crystal violet; demethylation; lignin; polyurethane coatings; triphenylmethane dyes

---

1. Introduction

Lignin, the most abundant natural resource next to cellulose and hemicellulose [1–4] contains various functional groups that provide active sites for chemical modification such as polarity adjustment to enhance the compatibility of lignin with other polymeric matrices in lignin/polymer composites [4,5] or to improve antioxidant properties [6–11]. Furthermore, studies reported lignin-derived encapsulation of various drugs for biomedical and agricultural applications. Richter et al. reported the encapsulation of silver nanoparticles in lignin-coated polymers [12]. Gregorova et al. studied the encapsulation of lignin nanoparticles in polyethylene films (Björkman lignin from beech wood flour) [13]. In other studies, the delivery of Resveratrol® [14], the controlled release of Avermectin® [15], lignin polyurea microcapsules with anti-photolysis and sustained-release performances [16], montmorillonite–lignin hybrid hydrogel as super-sorbent for dye removal from wastewater [17], cellulose–lignin hydrogels and their controlled release of polyphenols [18], lignin-stimulated protection of polypropylene films and DNA in cells of mice against oxidation damage [19] have been tested. Gao [20] and Bshena [21] studied the antimicrobial activity of various textiles, using lignin incorporated into polyethylene films and applied in the finishing processes. For textiles, there are special requirements such as non-toxicity to the consumer, namely cytotoxicity, allergy or irritation and sensitization. In other recent studies, lignosulfonic acid is reported to exhibit broad-spectrum anti-HIV (human immunodeficiency virus) and anti-HSV (herpes simplex virus) properties [22,23]. Thus, Qiu investigated the anti-HIV-1 activity-potential of lignosulfonates as a microbicide to prevent HIV-1 sexual transmission [23]. Another recently reported study revealed that the antimicrobial capacity of lignin correlates with the phenolic components, specifically the side chain structure and the nature of further functional groups [24]. Typically, the presence of a double bond in α, β positions of the side chain and a methyl group in the γ position grants the phenolic fragments with the most potency against microorganisms. However, none of the hitherto published studies included the investigation of the antibacterial activity of lignin when included in polymeric matrices.

Unmodified lignin is widely studied as a component for polymer production with a focus on phenol–formaldehyde resins and polyurethanes (PUs) [25], where lignin is used as polyol substitute due to the high amount of hydroxyl groups resulting in high crosslinking densities and variable mechanical properties [26–28]. In previous studies, lignin-derived polyurethane coatings have been prepared using Kraft lignin isolated at room temperature from aqueous media (black liquor) at different pH values [29]. In addition, their antioxidative activity has been investigated using the Folin–Ciocalteu (FC) assay [30]. Although lignin contains many functionalities, they are often difficult to access due to rather strong steric hindrance. So far, various procedures have been explored to incorporate more OH groups into the lignin structure including hydroxymethylation, phenolation, demethylation, oxidation and reduction [31]. These modifications have been studied primarily in conjunction with phenol–formaldehyde (PF) resins or PU research using lignin as a replacement for fossil-based phenols and polyols [32].

In 2016, Li et al. reported using demethylation to enhance the chemical reactivity at atmospheric pressure to produce fast curing phenolic resins [33]. Another possibility for lignin demethylation is an enzymatically catalyzed reaction using fungi (i.e., white and brown rot fungi) or bacteria (i.e., *Pseudomonas*, *Sphingomonas*). Mainly laccase was investigated, which oxidizes the guaiacyl into catechol units [34]. Industrially, demethylated lignin is recovered as a byproduct in dimethylsulfoxide (DMSO) production. For this purpose, black liquor is mixed with molten sulfur at about 230 °C. Two methyl groups are transferred from the lignin to the sulfur, forming dimethyl sulfide, which is oxidized to DMSO with nitrogen dioxide. Based on this process, Kraft lignin was demethylated with sulfur at 225 to 235 °C under high pressure and successfully increased its reactivity for the synthesis of phenol–formaldehyde (PF) resins [34]. Sulfur and halogen compounds are also used as nucleophiles for the chemical demethylation of lignin. For example, Chung and Washburn have demethylated softwood Kraft lignin with hydrobromic acid under the catalytic action of hexadecyltributylphosphonium bromide at 115 °C for 20 h, resulting in an increase in the OH content of 28% [35]. PU foams synthesized from the modified Kraft lignins showed a higher compressive strength than conventional ones [36,37]. Song et al. used the same method for white straw alkali lignin, with results that showed a significant increase in the total hydroxy content of demethylated lignin [38] when samples were explored for the synthesis of bio-based PF resins by demethylations with sulfur-containing compounds (sulfur, *n*-dodecyl mercaptan, sodium hydrogen sulfide and sodium sulfite). Here, soda lignin was heated with the reagent for 1 h at 90 °C. This research aimed to provide a cost-effective and efficient method for the chemical demethylation of lignin. The best results in terms of an increase in OH content and use for PF resins was the sample demethylated with Na_2SO_3. Other authors used Na_2SO_3 for demethylation performed under high-pressure reactors [39] or under reflux [15,40]. Podschun et al. chose a different approach in which organosolv lignin was demethylated under microwave radiation [41].

The antimicrobial properties of various dyes, in particular triphenylmethane (TPM) derivatives such as malachite green and crystal violet, have been studied since their first successful application as bioactive additives more than a hundred years ago (Figure 1, Table 1).

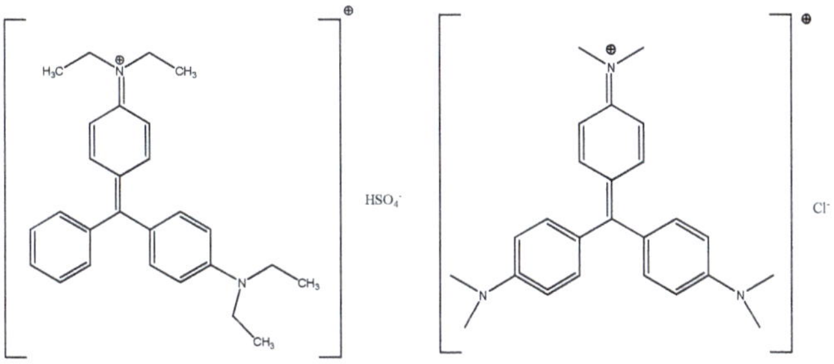

Figure 1. Molecular structure of (**a**) brilliant green and (**b**) crystal violet (see also Table 1).

In 1891, methylene blue, another TPM derivative was discovered by Paul Ehrlich to be efficient in malaria treatment, a few years later followed by the discovery of the antiseptic capacity of brilliant green (BG) [42–44]. For many years, malachite green was one of the most frequently used disinfectants in aquaculture due to its fungicidal effects. Due to the discovery of antibiotics and biocide polymers, antimicrobial dyes have not only been displaced in biomedicine but also other applications. Bolous et al. intensively studied the mechanisms of antimicrobial effects [45,46]. In detail, it was reported that the evidence to link the antimicrobial properties of TPM dyes, especially brilliant green, to the activity of mechanosensitive ion channel (MIC) of large conductance, which is known to be highly specific and ubiquitous in various bacterial species [47]. In 2012, Vilela et al. reported a study using methylene blue (MB) and malachite green photosensitizer microbial reduction of *Staphylococcus aureus* by synthesizing biofilms with it. The best results showed microbial reduction with 3000 μM of malachite green with a microbial reduction of 1.6–4.0 $\log_{10}$ [48]. In oral cavities, biofilm formation is considered to cause resistance to antimicrobial agents. Photodynamic therapies using phenothiazinic photosensitizers first confirmed the antimicrobial effect in biofilms. Malachite green was then compared with the phenothiazinic photosensitizers (methylene blue and toluidine blue) on *Staphylococcus aureus* and *Escherichia coli* biofilms. Noimark et al. reported the synthesis of modified photobactericidal silicones for medical applications. In detail, crystal violet and/or methylene blue were incorporated into the silicone bulk and gold nanoparticles were coated using a dipping method. The polymers showed good photostability, the photobactericidal activity was determined against *Staphylococcus epidermidis* and *Escherichia coli*. The results showed that these multi-dye–nanogold-polymers exhibit strong photobactericidal activity both under light and dark conditions [49]. Bartoszewicz et al. filed a patent claiming lubricious antimicrobial coatings containing silver, pyrrolidone carboxylic acid (PCA) and a TPM dye (malachite green). The coating composition of the invention provides photostability to the silver ions contained therein and is hydrophilic and antimicrobial [50]. In 2016, Santos et al. comprehensively reviewed various classes of antimicrobial polymers and discussed their bioactivity mechanisms including biocidal activity, antifungal and antibacterial capacity against numerous microorganisms (i.e., gram positive and gram negative bacteria and fungi) [51]. Table 1 summarizes literature reporting the antimicrobial activity of lignins and triphenylmethane derivatives such as malachite green, brilliant green, methylene blue and crystal violet.

Table 1. Literature studies regarding antimicrobial activity of lignins and triphenylmethane derivatives (i.e., malachite green, brilliant green, methylene blue and crystal violet).

Sample Composition	Studied Activity (Antibacterial, Antifungal)	Microorganisms (Bacteria, Fungi)	Results	References
Triphenylmethane (TPM) dyes	Mechanistic studies of the antimicrobial effects of triphenylmethanes (crystal violet, methylene blue, malachite green, brilliant green).	Various gram positive and gram negative bacteria	Evidence to link the antimicrobial properties of TPM dyes, especially brilliant green, to the activity of mechanosensitive ion channel (MIC) of large conductance, known to be highly specific/ubiquitous in various bacterial species.	Bolous et al. [45–47]
TPM dyes (i.e., methylene blue, malachite green) used as photosensitizer for acrylic resins	Microbial reduction of biofilms.	S. aureus	Best microbial reduction with 3000 µM malachite green with microbial reduction of 1.6–4.0 $\log_{10}$.	Vilela et al. [48]
TPM-based antimicrobial surfaces	Antimicrobial effects of crystal violet and methylene blue.	S. epidermidis (RP62a) and E. coli (NCTC 25522)	Light-activated antimicrobial surfaces with enhanced efficacy induced by a dark-activated mechanism.	Noimark et al. [49]
TPM-based coating additives (i.e., brilliant green, crystal violet) for polymeric substrates including PU	Antimicrobial activity of photo-stable composition used for coating a variety of medical materials.	Not specified	The coating composition comprising silver and TPM dyes (malachite green) provided photostability to the silver ions and antimicrobial activity.	Bartoszewicz et al. WO 2009/015476 Al [50]
Antimicrobial polymers	Bioactive polymers including biocidal activity, antifungal and antibacterial capacity.	Various gram positive and gram negative bacteria	Comprehensive review discussing different mechanisms regarding antimicrobial effects in polymer materials.	Santos et al. 2016 [51]
Lignin/HPMC and HPMC/lignin/chitosan composites	Antibacterial effects.	E. coli and S. aureus, B. thermosphacta and P. fluorescens	Testing the films against spoilage bacteria that grow at low temperatures revealed the activity of the 30% addition on HPMC/lignin against B. thermosphacta and P. fluorescens. HPMC/lignin/chitosan films (5% lignin) showed activity against both B. thermosphacta and P. fluorescens.	Alzagameem et al. 2019 [52]
Cellulose and lignin effects on disintegration, antimicrobial and antioxidant properties of PLA active films	Antimicrobial, antioxidant and disintegrability activities	Gram negative bacteria: Xanthomonas axonopodis pv. vesicatoria and Xanthomonas arboricola pv. pruni	Inhibition capacity for Gram negative bacteria (Xanthomonas axonopodis pv. vesicatoria and Xanthomonas arboricola pv. pruni) for lignin-modified PLA films.	Yang et al. 2016 [53]
Lignin derivatives (epoxides, esters, ether)	Antimicrobial activity of chemically modified lignins (by acetylation, epoxidation and hydroxymethylation reactions).	Bacillus aryabhattai and Klebsiella	Epoxy/lignin was found to be the most effective antibacterial among modified lignin with minimum inhibitory concentration of 90 and 200 µg/disc.	Kaur et al. 2017 [54]
Lignin for benign encapsulation	Antimicrobial activity of nanoparticles coated with LignoBoostTM softwood Kraft lignin.	E. coli and Pseudomonas aeruginosa	Nanoparticle flash precipitation with subsequent silver ion infusion and polyelectrolyte coating including lignin.	Richter et al. 2015 [12]
Antibacterial lignin–polyethylene (PE)	Lignin nanoparticles embedded in polyethylene films (Björkman lignin from beech wood flour).	E. coli and S. aureus	Lignin particles exhibit antibacterial effect against E. coli and S. aureus in the same order of magnitude as other antibacterial agents such as Bronopol® and Chlorohexidine®.	Gregorova et al. 2011 [13]

In a recently published study, chitosan/hydroxypropylmethylcellulose (HPMC) composites with varying ratio up to 30% of Kraft lignins (isolated from black liquor and purified via solvent extraction) were prepared and tested against spoilage bacteria that grow at low temperatures. The results revealed the activity against both *B. thermosphacta* and *P. fluorescens* for samples with 30% lignin. In HPMC/lignin/chitosan films, the 5% addition exhibited activity against both *B. thermosphacta* and *P. fluorescens* [52]. Currently, these lignin-derived composites are studied regarding their applications as scaffold component for mesenchymal stem cell differentiation and bone regeneration [55]. To do so, lignin as feedstock component has to be specified including protocols for quality control using novel chemometric data analysis methods [56,57].

In the present study, lignins isolated from black liquor at different pH values were used to explore the potential of these compounds as an antimicrobial component in polyurethane coatings. First, the extraction conditions that favored high lignin yields were optimized. Unmodified and demethylated lignins were used to prepare the lignin–polyurethane (LPU) coatings. The last part of the study aimed to correlate the antimicrobial properties with extraction conditions (i.e., pH value) and molecular structures (unmodified versus demethylated lignins). Furthermore, the influence of additional antimicrobial dyes (brilliant green and crystal violet) on the LPU coating bioactivity, color and morphology was studied.

2. Materials and Methods

2.1. Extraction of Kraft Lignin (KL) and Organosolv Lignin (OL)

The Kraft lignin (KL) was extracted through the acidic precipitation from black liquor according to a procedure reported by Garcia et al. [58]. First, about 450 mL of black liquor was filtered with a vacuum filter. The filter cake was rejected. Of the filtrate, 400 mL was heated to 50–60 °C. Sulfuric acid (160 mL, 25 vol.%) was added while stirring. The mixture was stirred for another hour at room temperature and then vacuum filtered. The filter cake was washed with distilled water and sulfuric acid (25 vol.%) until the requested pH value was reached (pH 2 to pH 5). Finally, the precipitated lignin was dried in a freeze dryer for 48 h. The organosolv lignin (OL) was isolated according to a procedure recently reported [10].

2.2. Synthesis of Demethylated Kraft Lignin

For the demethylation, a procedure reported by Li et al. was used and slightly modified [33]. The sample (1 g), 0.1 g of Na_2SO_3 as the demethylating reagent and 6 g of 2.5 mol NaOH solution were introduced into a 15 mL rolled rim glass on an analytical balance and homogenized. The solution was heated with stirring to 90 or 72 °C and stirred for 1 h at this temperature. After cooling to room temperature (RT), the pH was adjusted to pH 2 by means of 1% HCl. The demethylated lignin precipitated as a brown solid. The suspension was transferred to a 45 mL tube and centrifuged for 10 min at 3000 rpm to separate the demethylated lignin from the aqueous solution. The lignin was washed with distilled water and the pH adjusted to pH 7 with 2.5 molar NaOH solution. It was again centrifuged (for 30 min at 4000 rpm) to separate the aqueous phase from the demethylated lignin. The product was first stored at 40 °C in a drying oven and then freeze-dried at 80 °C and 0.10 mbar. Subsequently, the samples were homogenized and transferred for storage in rolled edge glasses, which were closed with snap lids. Furthermore, the samples were protected against UV radiation.

2.3. Size Exclusion Chromatography

Size exclusion chromatography was used to determine the number-average (M_n) and weight-average (M_w) molecular weights of lignins and their polydispersities, analogue to recently reported methods [29,30]. A PSS SECurity^2 GPC System was used with tetrahydrofuran as the mobile phase, a run time of 30 min and an injection volume of 100 μL. The system was calibrated using polystyrene standards at different molecular weights.

2.4. Determination of Hydroxyl Groups

The content of hydroxyl groups was determined via two different methods. ISO 14900:2001(E) developed for polyether polyols with steric hindrance was recently reported [29]. Shortly, each lignin sample was boiled under reflux in 25 mL of acetylation reagent solution with a blank sample simultaneously under the same conditions. After three hours at reflux, the flasks were left to cool down to room temperature. Twenty-five milliliters of sample and blank, respectively, were filled up with water to 100 mL and were titrated with sodium hydroxide (0.5 M). The split up of the acetylated samples and blanks allowed a triple determination via titration. Different amounts of sample and blank were needed. The differences were used to determine the total hydroxyl content.

2.5. Antibacterial Activity of Lignin

The antimicrobial activity of the lignin powders samples was analyzed in a quantitative way by modifying the test for antimicrobial activity and efficacy (JIS Z 2801:2010) of liquid samples [59]. The JIS is based on a comparison of bacteria counts in saline solution on reference and sample materials after a defined incubation temperature and time. *Staphylococcus aureus* (DSM No. 799) was applied as the test organism. The inoculum was prepared in the same way as described above. According to the McFarland-standard the inoculum was adjusted in physiological saline solution with tryptone (Blank, Vörstetten, Germany; VWR International, Darmstadt, Germany) to a concentration of 108 cfu mL^{-1}. This inoculum suspension was diluted in physiological saline solution with tryptone (Blank, Vörstetten, Germany; VWR International, Darmstadt, Germany) to a final concentration of 105 cfu mL^{-1}. Lignin powder was added into tubes with 5 mL physiological saline solution with tryptone to a final concentration of 0.1, 0.01 and 0.001 g mL^{-1}. Each tube was inoculated with 50 µL of the inoculum. The same measurements were done in nutrient broth instead of physiological saline solution. The measurements were carried out in triplicates.

The inoculum (1 mL) was incubated at 37 °C for 24 h in a mixture of 9 mL nutrient broth (Merck KGaA, Darmstadt, Germany) and 1 mL of sample or reference. Afterwards viable counts were determined by counting the colonies on plate-count agar after incubation at 37 °C for 24 h.

The value of antimicrobial activity was calculated by subtracting the logarithmic value of viable counts of the sample from the logarithmic value of reference material after inoculation and incubation:

$$log_{10} - reduction = log_{10}\left(\frac{c_{gew}(reference)}{c_{gew}(sample)}\right) \tag{1}$$

where as $c_{gew}(reference)$ = arithmetic mean of bacterial counts of reference 24 h after inoculation, and $c_{gew}(sample)$ = arithmetic mean of bacterial counts of sample material 24 h after inoculation. According to the JIS Z 2801:2010 a material can be characterized as antimicrobial, if the calculated log_{10}-reduction is ≥2.0 after 24 h at 37 °C [59].

2.6. Hemmhoff Test

The antimicrobial activity of the lignin was tested according to the disk diffusion test of the National Committee for Clinical Laboratory Standards (NCCLS) standard method. The disk diffusion test is based on the diffusion of the sampling material in agar. If the bacterium is sensitive to the tested substance, the growth of the bacterium is inhibited and a visible inhibition zone arises. The inhibition zone is the defined area between the punched out area and the beginning of the grown bacterium. If there is no inhibition zone, the bacterium is not sensitive to the tested substance.

Staphylococcus aureus (DSM No. 799) was used as a test organism. The inoculum was prepared by transferring a frozen culture to 10 mL of nutrient broth (Merck KGaA, Darmstadt, Germany). The nutrient broth with the inoculum was incubated at 37 °C for 24 h. According to the McFarland-standard the inoculum was adjusted in physiological saline solution with tryptone (Blank, Vörstetten, Germany; VWR International, Darmstadt, Germany) to a final concentration of 108 cfu mL^{-1}. In each Petri dish (Sarstedt AG, Nümbrecht, Germany) 100 µL of the inoculum was spatulated on Mueller–Hinton agar (VWR International, Darmstadt, Germany) which were impregnated with the different lignins and blank filter papers as references and were put on the inoculated agar plates.

The agar plates were incubated at 37 °C for 24 h. Afterwards, the diameter of the inhibition zone was measured with a digital caliper (Traceable Digital Caliper 6, VWR International, Darmstadt, Germany).

2.7. Synthesis of Lignin-Based Polyurethane Coatings

PEG400 was obtained from Sigma-Aldrich (Steinheim, Germany). 4,4-Diphenylmethane diisocyanate (MDI, for synthesis) was purchased from Merck in Darmstadt and triethylamine (TEA, for synthesis) was received from Carl Roth GmbH in Karlsruhe. All chemicals were used without further purification. PEG400 was mixed with lignin to obtain 1 g of polyol blend. Coatings prepared from lignins isolated at different pH values were produced analogously to the previously described procedure, with the MDI amount adapted to the hydroxyl number of the lignin and the resulting polyol blend. Lignin-based PU coatings were prepared using unmodified and demethylated lignins, respectively, and 4,4-diphenylmethandiisocyanate (MDI). The NCO:OH ratio was 1.7. The calculation was performed according to literature reference [60,61]:

$$\frac{\text{NCO}}{\text{OH}} = \frac{w_{\text{MDI}} \times [\text{NCO}]_{\text{MDI}}}{w_{\text{L}} \times [\text{OH}]_{\text{L}} + w_{\text{P}} \times [\text{OH}]_{\text{P}}} \tag{2}$$

where w_{MDI}, w_{L} and w_{P} are the weights (g) of MDI, lignin and polyol, respectively. $[\text{NCO}]_{\text{MDI}}$ is the molar content of isocyanate groups in MDI, 8.0 mmol/g for 4,4'-MDI. $[\text{OH}]_{\text{L}}$ and $[\text{OH}]_{\text{P}}$ are the molar contents of total hydroxyl groups in the lignin and the polyol, respectively. Masses of lignin and polyol were kept constant. Thus, 1 g of lignin was dissolved in 6 mL THF under constant stirring. MDI was added and the mixture was transferred on a polyethylene (PE) transparency and dried for 1 h at room temperature. Finally, the pre-films were cured at 37 °C for 3 h to obtain the final lignin PU films. The synthesis of lignin-modified PU coatings with brilliant green (BG) and crystal violet (CV) followed the same procedure, using 0.8% (w/v) of the corresponding triphenylmethane derivative.

Analogously, 1 g of demethylated lignin was dissolved in 6 mL of THF under constant stirring to prepare the LPU coatings. MDI was added and the mixture was transferred onto a PE-transparency and dried for 1 h at room temperature. Finally, the pre-films were cured at 35 °C for 3 h to obtain the final lignin-derived PU films.

2.8. Antimicrobial Activity of the LPU Coatings

The antimicrobial activity of the coatings was analyzed based on the Japanese Industrial Standard (JIS) Z 2801:2010 [59]. The JIS is based on the comparison of bacteria counts on sample coating/surface and reference material after a defined storage temperature and time (35 °C, 24 h). The reduction of bacteria counts were calculated and represented as $\log_{10}$-reduction. The $\log_{10}$-reduction is a measure for the antimicrobial activity and effectiveness of the coatings. According to the JIS a material is called antimicrobial when the $\log_{10}$-reduction is $\geq 2 \log_{10}$.

Staphylococcus aureus (DSM No. 799) and *Listeria monocytogenes* were used as test organisms. The inoculum was prepared by transferring a frozen culture to 10 mL of nutrient broth (Merck KGaA, Darmstadt, Germany). The nutrient broth with the inoculum was incubated at 37 °C for 24 h. According to the McFarland-standard the inoculum was adjusted in physiological saline solution with tryptone (Blank, Vörstetten, Germany; VWR International, Darmstadt, Germany) to a final concentration of 108 cfu mL^{-1}. This inoculum suspension was diluted in physiological saline solution with tryptone to a final concentration of 105 cfu mL^{-1}.

The coatings and references were inoculated with 400 µL of the inoculum suspension. To enlarge the contact area of the coatings with the inoculum, the inoculum was covered with a sterile foil (Interscience,

Saint-Nom-la-Bretèche, France). The plates were incubated at 37 °C for 24 h. After incubation the inoculated suspension was washed out with 10 mL soybean casein lecithin polysorbate 80 broth (SCDLP) solution (Merck KGaA, Darmstadt, Germany). This served as the first solution stage and was used for further decimal solution series. The bacteria counts were determined by using the drop-plate-technique and counting the colonies on plate-count agar (Merck KGaA, Darmstadt, Germany) after incubation at 37 °C for 24 h.

The value of antimicrobial activity was calculated by subtracting the logarithmic value of viable counts of the sample from the logarithmic value of reference material after inoculation and incubation:

$$log_{10} - reduction = log_{10}(\frac{c_{gew}(reference)}{c_{gew}(sample)}) \tag{3}$$

where c_{gew} *(reference)* = arithmetic mean of bacterial counts of reference 24 h after inoculation, and $c_{gew}(sample)$ = arithmetic mean of bacterial counts of sample material 24 h after inoculation. According to the JIS Z 2801:2010 a material can be characterized as antimicrobial, if the calculated log_{10}-reduction is ≥2.0 after 24 h at 37 °C.

2.9. Thermogravimetric Analysis

TGA measurements were performed with about 10 mg of lignin using a Netzsch (Selb, Germany) TGA 209 F1 with a heating rate of 10 °C min^{-1} under a nitrogen atmosphere. The temperature ranged from ambient to 800 °C.

2.10. Optical Contact Angle

Static optical contact angle (OCA) measurements were performed on the PU films at room temperature using an OCA device equipped with a charge-coupled device (CCD) photocamera (DataPhysics Instruments, Filderstadt, Germany). A 40 µL volume of distilled water was used to dispense liquid droplets.

2.11. Scanning Electron Microscopy

Scanning electron microscopy (SEM) from ThermoFischer was combined with X-ray analysis (SEM-EDX). Characterization of the texture, phases and the thin LPU layer were determined by SEM-EDX microscopy using an ESEM Quanta FEG 250 FEI with Apollo XL30 EDX (Thermo Fisher Scientific Inc., Huntsville, AL, USA).

3. Results and Discussion

3.1. Antibacterial Activity of Kraft Lignin

Kraft lignins were demethylated (DL) and characterized regarding their molecular weight and hydroxyl content (Table 2). In addition, Table 2 shows Kraft lignins isolated at different pH values [29].

Studies of the antimicrobial activity were performed following procedures reported to investigate intrinsically antimicrobial polymers based on poly((tertbutyl-amino)-methyl-styrene) [62–66] and coatings based on HPMC/lignin/chitosan [52]. Two different nutritions were used: sodium chloride (NaCl) and physiological saline solution (NB) of different concentrations (Figure 2).

Results for both solutions (NaCl, NB) clearly showed an increase in antimicrobial activity against *Staphylococcus aureus* for the lignins isolated at different pH values (pH 2 to pH 5) with the highest activity (log_{10} reduction of 7.0) for the pH 5 samples. This tendency could also be confirmed for the corresponding LPU coatings prepared from the different lignin samples (see next paragraph). Due to the measurement procedure, the study started using the highest concentrations (0.1 mol/L), then the concentration decreased down to 0.001 mol/L. Obviously, the lowest concentrations were sufficient for the observed antimicrobial effects. Similar results could be observed for the HPMC/lignin coatings recently reported [52] and also for organosolv lignins (not yet published). Further studies are required to clarify the correlation of concentration and antimicrobial activity.

Table 2. Weight-average (M_w) and number-average (M_n) molecular weight and polydispersity (PDI) obtained by gel permeation chromatography (GPC) measurements, and OH content according to ISO 14900 for demethylated lignins (DL) and Kraft lignins isolated at different pH values [29].

Lignin	M_w (g/mol)	M_n (g/mol)	PDI	OH content (ISO 14900) ($mmol·g^{-1}$)	(mg KOH) g^{-1}	Reference
pH2	1879	574	3.3	2.67	150	[29]
pH3	1732	538	3.2	4.48	251	[29]
pH4	1570	441	3.6	5.02	282	[29]
pH5	1502	490	3.0	5.34	300	[29]
DL-pH2	5417	1299	4.2	4.75	266	–
DL-pH3	5461	1318	4.1	4.00	224	–
DL-pH4	5522	1335	4.1	5.51	309	–
DL-pH5	5610	1347	4.2	4.80	269	–

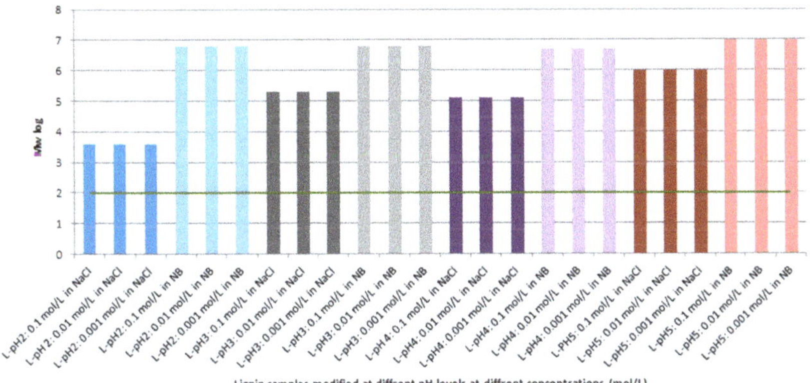

Figure 2. Antimicrobial activity of unmodified Kraft lignins isolated at different pH levels (varying from 2 to 5). Activity tested against *Staphylococcus aureus* in NaCl and physiological saline solution (NB)), respectively, in concentrations ranging between 0.1–0.001 mol/L.

3.2. Antibacterial Activity of LPU Coatings

For the comparability of subsequent investigations, first the antimicrobial effect on different reference surfaces was tested according to Japanese Industrial Standard Z 2801:2000 [59]. The results are shown in Table 3 and Figure 3.

Table 3. Results of antimicrobial activity of different reference systems surfaces against *S. aureus*.

Reference Systems (blank)	Kbe mL^{-1}	Ø log Kbe mL^{-1}
Petri dish	1.18×10^5	5.05
Glass	1.07×10^7	6.71
Plastic dish (PP *)	4.07×10^6	6.48
Transparencies (PS **)	9.62×10^6	6.94
Stainless steel	2.60×10^1	1.41

* Polypropylene, ** Polystyrene

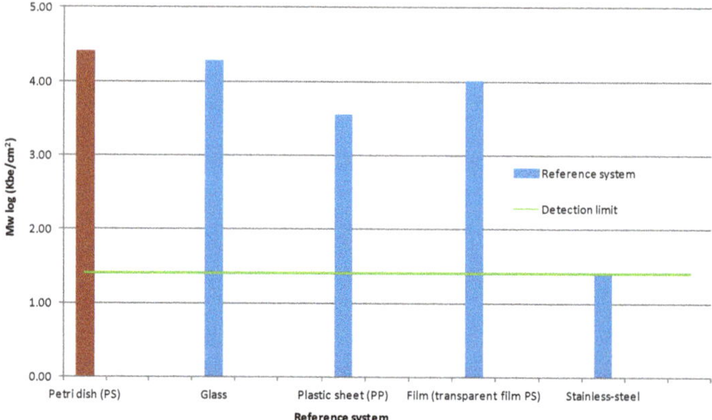

Figure 3. Antimicrobial activity (log Kbe mL^{-1}) of different blank surfaces (used as reference systems) against *S. aureus* and *Listeria monocytogenes*. The green line represents the detection limit for the determination of the antimicrobial activity and is 1.4 log (Kbe cm^{-1}).

As suggested, the results showed normal bacterial growth on blank surfaces: untreated glass, plastic sheets (polypropylene, PP), transparent polystyrene films (PS) and stainless steel (Figure 3). Notable was an emerging germ resistance on the untreated stainless-steel surfaces which underlines the natural antimicrobial effect of stainless-steel surfaces for different bacteria, known as oligodynamic effect. The oligodynamic effects describes the damaging effect of various metal ions on different bacteria, viruses and fungi, most probably due stainless-steel alloy formation initiated by different metal cations [64].

Furthermore, lignin-modified PU coatings prepared from demethylated lignins were applied to various surfaces and analyzed for their antimicrobial action. The results are listed below in Table 4 and Figure 4.

Table 4. Results of antimicrobial activity of demethylated lignin-based polyurethane (PU) coatings against *S. aureus*.

Lignin–PU Coatings	Kbe/cm^2	Ø log Kbe/cm^2
DL-pH2-060718	2.07×10^3	3.03
DL-pH3-060718	1.17×10^3	2.51
DL-pH4-060718	3.09×10^1	2.36
DL-pH5-060718	6.25×10^{-1}	0

The results showed significant microbial reduction against *S. aureus* for the PU coatings synthesized from demethylated lignins. It is also noticeable that the germ reduction can be correlated to the pH value for lignin isolation: lignins isolated at pH 3, 4 and 5 showed a higher germ reduction and antimicrobial activity, respectively, than the reference (blind value: polypropylene glycol (PPG) as polyol without lignin). One reason for this could be the improved homogeneity of the coatings, caused by higher crosslinking density of the LPU due to high OH numbers, analogous to the correlation of OH number and antioxidant activity of LPU coatings [29,30]. Besides LPU coatings, it was recently reported that the antimicrobial activity of various lignin-derived cellulose and cellulose/chitosan composites against *S. aureus* and *E. coli* (Table 1) [52]. A comparison of hydroxypropylmethyl cellulose/lignin films were blended with Kraft lignin in different amounts up to 30 wt.%. Comparing both systems (HPMC versus PU), the capacity against *S. aureus* was highest for the lignin isolated at pH 5 (Table 5). As supposed, the addition of triphenylmethane derivatives (BG, crystal violet (CV)) resulted in increased antimicrobial activity against *S. aureus*.

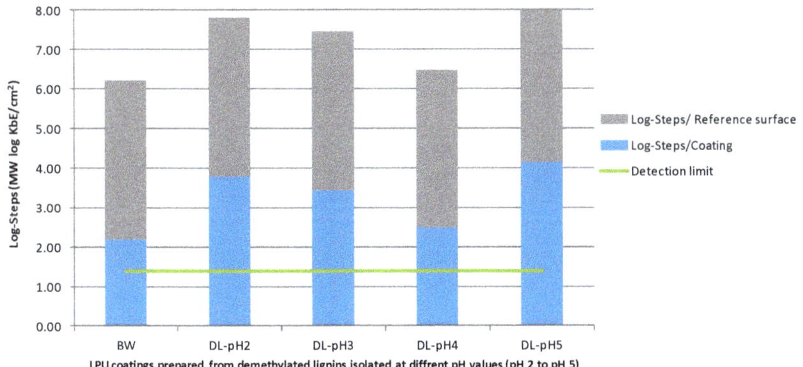

Figure 4. Results of antimicrobial activity against *S. aureus* of lignin-based PU coatings prepared demethylated lignins (DL) isolated at different pH values ranging from pH 2 to pH 5. BV:blind value of PU without lignin. The green line represents the detection limit for the determination of the antimicrobial activity (1.4 log Kbe cm^{-1}).

Table 5. Antimicrobial activity of lignin and lignin-derived PU coatings against *S. aureus*. The lignins used for lignin–polyurethane (LPU) coating preparation were isolated from aqueous solution at different pH values. For comparison, the antimicrobial activity of hydroxypropylmethylcellulose (HPMC)/lignin coatings was added in this table, previously reported in [52].

Antimicrobial Activity	Lignin (Isolated at pH 5)	LPU Coating (DL-pH5)	LPU Coating (KL-pH5)	LPU with 0.8% (*w/v*) BG	LPU with 0.8% (*w/v*) CV	HPMC/lignin (15 wt.% L1) [52]
Log$_{10}$ reduction	7.00	4.12	2.62	8.31	8.60	2.50

3.3. Thermal Properties (TGA)

TGA measurements were performed to describe and evaluate the thermal stability of the corresponding LPU coatings with antimicrobial additives and the coatings prepared from demethylated lignins (Figure 5).

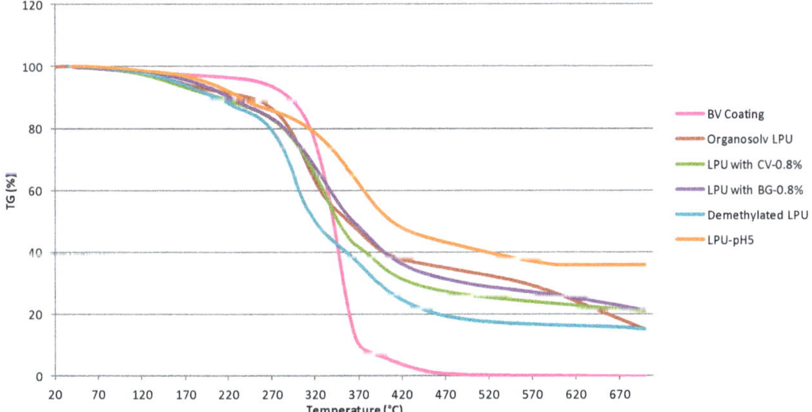

Figure 5. TGA results of different modified lignin-based LPU coatings containing brilliant green (BG) and crystal violet (CV); BV (blind value: PU without lignin).

TGA results showed thermal stability between 143–165 °C which are reasonable temperature stabilities for applications in construction and packaging (Table 6). Coatings with CV as antimicrobial additive showed the highest temperature stability with 165 °C in contrast to LPU coatings prepared with BG with a stability of 146 °C. The residual mass of both LPU coatings was between 20%–21.5%.

Table 6. Thermal stability for lignin and various LPU coatings with/without triphenylmethane (TPM) derivatives.

Lignin Coatings	Temperature (°C)	Δm (%)	Residual Mass (%)
Blank (PU coating without lignin)	250	−4.75%	0.39
LPU-pH 5	166	−3.31%	35.99
Lignin coating CV	165	−4.05%	20.80
Lignin coating BG	146	−4.49%	21.43
Lignin coating organosolv	143	−4.03%	8.00
Lignin-DLPU coating	153	−4.55%	21.43

Obviously, decomposition temperature and residuals are influenced not only by the pulping process used for lignin isolation (Kraft versus organosolv), but also by demethylation and added antimicrobial triphenylmethane derivatives (brilliant green and crystal violet). Further quantification of these effects by DSC measurements is under investigation.

3.4. Contact Angle of LPU Coatings

The wettability properties of the surfaces of all lignin-based PUs were investigated by means of static contact angle measurements against water (Table 6). The LPU coatings possessed a rather hydrophobic character with water contact angles θH_2O up to 92 degrees, higher than literature data for LPUs reported by Jia et al. prepared from organosolv lignin (61°). The PU coatings with demethylated lignins showed a contact angle of 84.22 ± 0.51°. Table 7 summarizes the contact angle data of all LPU with antimicrobial additives. The results revealed that the LPU with brilliant green had a better wettability (87.36 ± 0.15°) compared to the LPU with crystal violet (67.40 ± 0.18°).

Table 7. Results of contact angle measurements of the different LPU. Abbreviations: polyurethane (PU), Kraft lignin (KL), organosolv lignin (OL), beech wood (BW), brillant green (BG), crystal violet (CV).

Sample	Contact Angle (°)
PU-KL-pH 2	92.28 ± 0.49
PU-KL-pH 3	80.49 ± 1.03
PU-KL-pH 4	83.28 ± 0.24
PU-KL-pH 5	86.01 ± 0.22
PU-OL	61,59 ± 0.69
PU-KL-Demethylated	84.22 ± 0.51
LPU Coatings with TPM dyes	
PU-BV-BG	62.93 ± 0.34
PU-BV-CV	80.19 ± 0.28
PU-KL-pH2-BG	87.36± 0.15
PU-KL-pH2-CV	81.11 ± 0.18

3.5. Morphology of the LPU Coatings

To get a first idea of the homogeneity of the lignin-derived coatings, the coatings were observed via reflected light microscopy showing that homogeneous coatings could be obtained using lignins of number-average molecular weight (M_n) < 500 g/mol (equivalent weight-average (M_w) < 1570 g/mol) and OH content above 5 mmol/g (samples isolated at pH 4 and pH 5). Using scanning electron microscopy (SEM), the thickness of the casted films was determined to range between 150–160 µm [29].

Figure 6 shows the prepared LPU coatings: (a) with 0.8% brilliant green added resulted in homogeneous films of greenish color and smooth surface; (b) with 0.8% crystal violet added also resulted in homogeneous films of smooth surface colored in dark blue. Reference coatings are shown containing PU/CV (no lignin) and PU (no lignin, no crystal violet). The antimicrobial activities were determined according to Japanese Industrial Standard (JIS) Z 2801:2010 [59]. Figure 6c shows coatings on different surfaces: stainless steel, wood, plastic (polypropylene). On all surfaces, the coatings showed smooth homogeneous surfaces. In ongoing studies, the adhesion strength will be quantified.

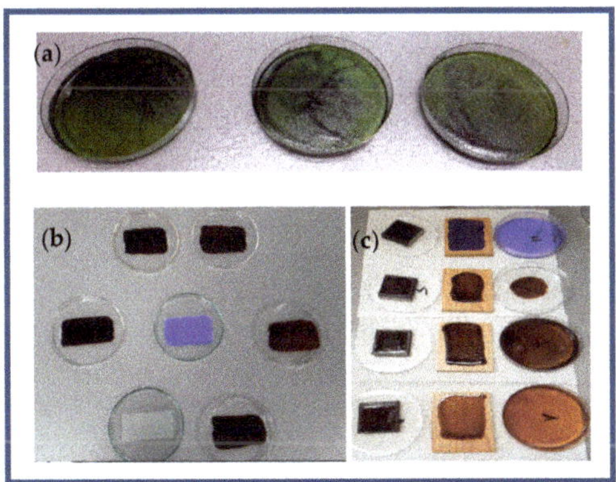

Figure 6. (a) Three lignin–PU coatings with brilliant green as antimicrobial additive; (b) lignin–PU coatings with crystal violet as an antimicrobial additive (blue sample in the middle: PU-CV without lignin; clockwise starting with the white sample (PU without lignin), LPU coatings prepared from organosolv-lignin and lignins isolated at pH 2 to pH 5; (c) lignin–PU–CV coatings on different surfaces: steel (left), wood (middle) and polystyrene (PS) petri dishes (right); top-down: PU-CV, LPU-CV-pH5, LPU-CV-pH4, LPU-CV-pH2.

4. Conclusions

The results of the antimicrobial activity study of lignin-based polyurethane coatings confirmed the capacity of Kraft lignin against special microorganisms such as *S. aureus*. Triphenylmethane derivatives (brilliant green, crystal violet) significantly increased this antimicrobial effect. The coating color changed from dark brown to green (in case of BG) and blue (in case of CV). Wettability tests using contact angle measurements confirmed the hydrophobic character of the lignin-derived PU coatings.

Author Contributions: S.E.K. mainly contributed to the manuscript, performed the experiments and analyzed the data; A.A. contributed analytical data regarding the lignins extracted from organic solvents; J.R. contributed in PU coating preparation. I.K. and J.K. contributed antimicrobial analyses; M.S. conceived and designed the experimental studies and contributed in writing the manuscript.

Funding: This research was supported by the Federal Ministry of Education and Research (BMBF) program "IngenieurNachwuchs" project LignoBau (03FH013IX4) and EFRE Infrastrukturförderung "Biobasierte Produkte" (EFRE0500035); Bonn-Rhein-Sieg University/Graduate Institute and Erasmus-Mundus Avempace-II scholarship (A.A.); and Bonn-Rhein-Sieg University/TREE institute (S.E.K.).

Acknowledgments: The authors gratefully acknowledge Zellstoff- und Papierfabrik Rosenthal GmbH (Blankenstein, Germany, MERCER group) for providing the black liquor. We thank Michael Larkins (Raleigh, North Carolina, U.S.) for final English corrections.

Conflicts of Interest: There are no conflicts to declare.

References

1. Ralph, J.; Lapierre, C.; Boerjan, W. Lignin structure and its engineering. *Curr. Opin. Biotechnol.* **2019**, *56*, 240–249. [CrossRef] [PubMed]
2. Rinaldi, R.; Jastrzebski, R.; Clough, M.T.; Ralph, J.; Kennema, M.; Bruijnincx, P.C.A.; Weckhuysen, B.M. Paving the way for lignin valorisation: Recent advances in bioengineering, biorefining and catalysis. *Angew. Chem. Int. Ed.* **2016**, *55*, 2–54. [CrossRef] [PubMed]

3. Alzagameem, A.; El Khaldi-Hansen, B.; Kamm, B.; Schulze, M. Lignocellulosic biomass for energy, biofuels, biomaterials, and chemicals. In *Biomass and Green Chemistry*, 1st ed.; Vaz, S., Jr., Ed.; Springer International Publishing: Basel, Switzerland, 2018; pp. 95–132.

4. Hansen, B.; Kamm, B.; Schulze, M. Qualitative and quantitative analysis of lignins from different sources and isolation methods for an application as a biobased chemical resource and polymeric material. In *Analytical Techniques and Methods for Biomass Products*; Vaz, S., Jr., Seidl, P., Eds.; Springer: Berlin, Germany, 2017; pp. 15–44.

5. Naseem, A.; Tabasum, S.; Zia, K.M.; Zuber, M.; Ali, M.; Noreen, A. Lignin-derivatives based polymers, blends and composites: A review. *Int. J. Biol. Macromol.* **2016**, *93*, 296–313. [CrossRef] [PubMed]

6. Ko, F.K.; Goudarzi, A.; Lin, L.-T.; Li, Y.; Kadla, J.F. Lignin-based composite carbon nanofibers. In *Lignin in Polymer Composites*, 1st ed.; Faruk, O., Sain, M., Eds.; Elsevier B.V: Amsterdam, The Netherlands, 2016; pp. 167–194.

7. Ponomarenko, J.; Dizhbite, T.; Lauberts, M.; Viksna, A.; Dobele, G.; Bikovens, O.; Telysheva, G. Characterization of softwood and hardwood lignoboost kraft lignins with emphasis on their antioxidant activity. *Bioresources* **2014**, *9*, 2051–2068. [CrossRef]

8. Benzie, I.F.; Devaki, M. The ferric reducing/antioxidant power (FRAP) assay for non-enzymatic antioxidant capacity: Concepts, procedures, limitations and applications. In *Measurement of Antioxidant Activity & Capacity*, 1st ed.; Apak, R., Capanoglu, E., Shahidi, F., Eds.; John Wiley & Sons Ltd.: Hoboken, NJ, USA, 2017; pp. 77–106.

9. Alzagameem, A.; El Khaldi-Hansen, B.; Büchner, D.; Larkins, M.; Kamm, B.; Witzleben, S.; Schulze, M. Lignocellulosic biomass as source for lignin-based environmentally benign antioxidants. *Molecules* **2018**, *23*, 2664. [CrossRef] [PubMed]

10. Bergs, M.; Völkering, G.; Kraska, T.; Do, X.; Monakhova, Y.; Konow, C.; Pude, R.; Schulze, M. *Miscanthus x giganteus* stem versus leave-derived lignins differing in monolignol ratio and linkage. *Int. J. Mol. Sci.* **2019**, *20*, 1200. [CrossRef]

11. Hansen, B.; Kamm, B.; Schulze, M. Qualitative and quantitative analysis of lignin produced from beech wood by different conditions of the Organosolv process. *J. Polym. Environ.* **2016**, *24*, 85. [CrossRef]

12. Richter, A.P.; Brown, J.S.; Bharti, B.; Wang, A.; Gangwal, S.; Houck, K.; Cohen Hubal, E.A.; Paunov, V.N.; Stoyanov, S.D.; Velev, O.D. An environmentally benign antimicrobial nanoparticle based on a silver-infused lignin core. *Nat. Nanotechnol.* **2015**, *10*, 817–824. [CrossRef]

13. Gregorova, A.; Redik, S.; Sedlarik, V.; Stelzer, F. Lignin-containing polyethylene films with antibacterial activity. In Proceedings of the 3rd International Conference on Thomson Reuters of NANOCON, Brno, Czech Republic, 21–23 September 2011. Available online: http://konference.tanger.cz/data/nanocon2011/sbornik/lists/papers/1366.pdf (accessed on 5 May 2019).

14. Dai, L.; Liu, R.; Hu, L.-Q.; Zou, Z.-F.; Si, C.-L. Lignin nanoparticle as a novel green carrier for the efficient delivery of resveratrol. *ACS Sustain. Chem. Eng.* **2017**, *5*, 8241–8249. [CrossRef]

15. Li, Y.; Yang, D.; Lu, S.; Lao, S.; Qiu, X. Modified lignin with anionic surfactant and its application in controlled release of avermectin. *J. Agric. Food Chem.* **2018**, *66*, 3457–3464. [CrossRef]

16. Pang, Y.; Li, X.; Wang, S.; Qiu, X.; Yang, D.; Lou, H. Lignin-polyurea microcapsules with anti-photolysis and sustained-release performances synthesized via pickering emulsion template. *React. Funct. Polym.* **2018**, *123*, 115–121. [CrossRef]

17. Wang, Y.; Xiong, Y.; Wang, J.; Zhang, X. Ultrasonic-assisted fabrication of montmorillonite-lignin hybrid hydrogel: Highly efficient swelling behaviors and super-sorbent for dye removal from wastewater. *Colloids Surf. A Physicochem. Eng. Asp.* **2017**, *520*, 903–913. [CrossRef]

18. Ciolacu, D.; Oprea, A.M.; Anghel, N.; Cazacu, G.; Cazacu, M. New cellulose–lignin hydrogels and their application in controlled release of polyphenols. *Mater. Sci. Eng. C* **2012**, *32*, 452–463. [CrossRef]

19. Kosikova, B.; Labaj, J. Lignin-stimulated protection of polypropylene films and DNA in cells of mice against oxidation damage. *Bioresources* **2009**, *4*, 805–815.

20. Gao, Y.; Cranston, R. Recent advances in antimicrobial treatments of textiles. *Text. Res. J.* **2008**, *78*, 60–72. [CrossRef]

21. Bshena, O.; Heunis, T.D.; Dicks, L.M.; Klumperman, B. Antimicrobial fibers: Therapeutic possibilities and recent advances. *Future Med. Chem.* **2011**, *3*, 1821–1847. [CrossRef]

22. Gordts, S.C.; Férir, G.; D'huys, T.; Petrova, M.I.; Lebeer, S.; Snoeck, R.; Andrei, G.; Schols, D. The low-cost compound lignosulfonic acid (LA) exhibits broad-spectrum anti-HIV and anti-HSV activity and has potential for microbicidal applications. *PLoS ONE* **2015**, *10*, e0131219. [CrossRef]

23. Qiu, M.; Wang, Q.; Chu, Y.; Yuan, Z.; Song, H.; Chen, Z.; Wu, Z. Lignosulfonic acid exhibits broadly anti-HIV-1activity-potential as a microbicide candidate for the prevention of HIV-1 sexual transmission. *PLoS ONE* **2012**, *7*, e35906. [CrossRef]

24. Kai, D.; Tan, M.J.; Chee, P.L.; Chua, Y.K.; Yap, Y.L.; Loh, X.J. Towards lignin-based functional materials in a sustainable world. *Green Chem.* **2016**, *18*, 1175–1200. [CrossRef]

25. Lau, P.C.K. Lignin: A platform for renewable aromatic polymeric materials. In *Quality Living Through Chemurgy and Green Chemistry. Green Chemistry and Sustainable Technology*; Lau, P.C.K. Springer: Berlin/Heidelberg, Germany, 2016; pp. 221–263.

26. Ten, E.; Vermerris, W. Recent developments in polymers derived from industrial lignin. *J. Appl. Polym. Sci.* **2015**, *132*, 1–13. [CrossRef]

27. Jia, Z.; Lu, C.; Zhou, P.; Wang, L. Preparation and characterization of high boiling solvent lignin-based polyurethane film with lignin as the only hydroxyl group provider. *RSC Adv.* **2015**, *5*, 53949–53955. [CrossRef]

28. Griffini, G.; Passoni, V.; Suriano, R.; Levi, M.; Turri, S. Polyurethane coatings based on chemically unmodified fractionated lignin. *ACS Sustain. Chem. Eng.* **2015**, *3*, 1145–1154. [CrossRef]

29. Klein, S.E.; Rumpf, J.; Kusch, P.; Albach, R.; Rehahn, M.; Witzleben, S.; Schulze, M. Utilization of unmodified kraft lignin for the preparation of highly flexible and transparent polyurethane coatings. *RSC Adv.* **2018**, *8*, 40765. [CrossRef]

30. Klein, S.E.; Rumpf, J.; Rehahn, M.; Witzleben, S.; Schulze, M. Biobased flexible polyurethane coatings prepared from kraft lignin: One-pot synthesis and antioxidant activity. *J. Coat. Technol. Res.* **2019**. [CrossRef]

31. Hu, J.; Zhang, Q.; Lee, D.-J. Kraft lignin biorefinery: A proposal. *Bioresour. Technol.* **2017**, *247*, 1181–1183. [CrossRef]

32. Sain, M.; Faruk, O. *Lignin in Polymer Composites*, 1st ed.; Elsevier: Kidlington, UK, 2016.

33. Li, J.; Wang, W.; Shifeng, Z.; Qiang, G.; Zhang, W.; Li, J. Preparation and characterization of lignin demethylated at atmospheric pressure and its application in fast curing biobased phenolic resins. *RSC Adv.* **2016**, *6*, 67435–67443. [CrossRef]

34. Laurichesse, S.; Avérous, L. Chemical modification of lignins: Towards biobased polymers. *Prog. Polym. Sci.* **2014**, *39*, 1266–1290. [CrossRef]

35. Chung, H.; Washburn, N.R. Improved lignin polyurethane properties with Lewis acid treatment. *ACS Appl. Mater. Interfaces* **2012**, *4*, 2840–2846. [CrossRef]

36. Zou, L.; Ross, B.M.; Hutchison, L.J.; Christopher, L.P.; Dekker, R.F.; Malek, L. Fungal demethylation of Kraft lignin. *Enzyme Microb. Technol.* **2015**, *73–74*, 44–50. [CrossRef]

37. Ibrahim, V.; Mendoza, L.; Mamo, G.; Hatti-Kaul, R. Blue laccase from *Galerina* sp.: Properties and potential for Kraft lignin demethylation. *Process Biochem.* **2011**, *46*, 379–384. [CrossRef]

38. Song, Y.; Wang, Z.; Yan, N.; Zhang, R.; Li, J. Demethylation of wheat straw alkali lignin for application in phenol formaldehyde adhesives. *Polymers* **2016**, *8*, 209. [CrossRef]

39. An, X.; Schroeder, H.A.; Thompson, G.E. Demethylated kraft lignin as a substitute for phenol in wood adhesive. *Chem. Ind. For. Prod.* **1995**, *15*, 36–42.

40. Ferhan, M.; Sain, M.; Yan, N. A new method for demethylation of lignin from woody biomass using biophysical methods. *J. Chem. Eng. Process. Technol.* **2013**, *4*, 160. [CrossRef]

41. Podschun, J.; Saake, B.; Lehnen, R. Catalytic demethylation of organosolv lignin in aqueous medium using indium triflate under microwave irradiation. *React. Funct. Polym.* **2017**, *119*, 82–86. [CrossRef]

42. Webb, C.H.S. A note on the value of brilliant green as an antiseptic. *Br. Med. J.* **1917**, *1*, 870. [CrossRef]

43. Sneader, W. *Drug Discovery: A history*; John Wiley and Sons Ltd.: Chichester, UK, 2005; p. 468.

44. Schirmer, R.H.; Coulibaly, B.; Stich, A.; Scheiwein, M.; Merkle, H.; Eubel, J.; Becker, K.; Becher, H.; Müller, O.; Zich, T.; et al. Methylene blue as an antimalarial agent. *Redox Rep.* **2003**, *8*, 272–275. [CrossRef]

45. Boulos, R. Bacterial Mechanosensitive Channels as Novel Targets for Antibacterial Agents. Ph.D. Thesis, The University of Western Australia, Perth, Australia, December 2011.

46. Boulos, R.A. Antimicrobial Compounds. U.S. Patent 20120329871 A1, 27 December 2012.

47. Boulos, R.A.; Eroglu, E.; Chen, X.; Scaffidi, A.; Edwards, B.R.; Toster, J.; Raston, C.L. Unravelling the structure and function of human hair. *Green Chem.* **2013**, *15*, 1268–1273. [CrossRef]

48. Vilela, S.F.G.; Junqueira, J.C.; Barbosa, J.O.; Majewski, M.; Munin, E.; Jorge, A.O.C. Photodynamic inactivation of *Staphylococcus aureus* and *Escherichia coli* biofilms by malachite green and phenothiazine dyes: An in vitro study. *Arch. Oral Biol.* **2012**, *57*, 704–710. [CrossRef]

49. Noimark, S.; Allan, E.; Parkin, I.P. Light-activated antimicrobial surfaces with enhanced efficacy induced by a dark-activated mechanism. *Chem. Sci.* **2014**, *5*, 2216. [CrossRef]

50. Bartoszewicz, L. Antimicrobial Photo-Stable Coating Composition. WO2009015476A1, 5 February 2009.

51. Santos, M.R.E.; Fonseca, A.C.; Mendonça, P.V.; Branco, R.; Serra, A.C.; Morais, P.V.; Coelho, J.F.J. Recent developments in antimicrobial polymers: A review. *Materials* **2016**, *9*, 599. [CrossRef]

52. Alzagameem, A.; Klein, S.E.; Bergs, M.; Do, X.T.; Korte, I.; Dohlen, S.; Kreyenschmidt, J.; Kamm, B.; Larkins, M.; Schulze, M. Antimicrobial activity of lignin and lignin-derived cellulose and chitosan composites against selected pathoge nic and spoilage microorganisms. *Polymers* **2019**, *11*, 670. [CrossRef]

53. Yang, W.; Fortunati, E.; Dominici, F.; Giovanale, G.; Mazzaglia, A.; Balestra, G.M.; Kenny, J.M.; Puglia, D. Effect of cellulose and lignin on disintegration, antimicrobial and antioxidant properties of PLA active films. *Int. J. Biol. Macromol.* **2016**. [CrossRef]

54. Kaur, R.; Uppal, S.K.; Sharma, P. Antioxidant and antibacterial activities of sugarcane bagasse lignin and chemically modified lignins. *Sugar Tech.* **2017**, *19*, 675–680. [CrossRef]

55. Witzler, M.; Alzagameem, A.; Bergs, M.; El Khaldi-Hansen, B.; Klein, S.E.; Hielscher, D.; Kamm, B.; Kreyenschmidt, J.; Tobiasch, E.; Schulze, M. Lignin-derived biomaterials for drug release and tissue engineering. *Molecules* **2018**, *23*, 1885. [CrossRef]

56. Monakhova, Y.; Diehl, B.W.K.; Do, X.T.; Witzleben, S.; Schulze, M. Novel method for the determination of average molecular weight of natural polymers based on 2D DOSY NMR and chemometrics: Example of heparin. *J. Pharm. Biomed. Anal.* **2018**, *149*, 128–132. [CrossRef]

57. Alzagameem, A.; Bergs, M.; Do, X.T.; Klein, S.E.; Rumpf, J.; Larkins, M.; Monakhova, Y.; Pude, R.; Schulze, M. Low-input crops as lignocellulosic feedstock for second generation biorefineries and the potential of chemometrics in biomass quality control. *Appl. Sci.* **2019**, *9*, 2252. [CrossRef]

58. Garcia, A.; Toledano, A.; Serrano, A.; Egüés, I.; González, M.; Marín, F.; Labidi, J. Characterization of lignins obtained by selective precipitation. *Sep. Purif. Technol.* **2009**, *68*, 193–198. [CrossRef]

59. Japanese Industrial Standard. Z 2801:2000. ICS 07.100.10; 11.100 Descriptors: Bacteriocide-Activity Determination, Microbiological-Resistance Tests, Biological Hazards, Culture Techniques. Available online: http://lotusyapi.com.tr/Antibacterial/JIS%20Z%202801%202000.pdf (accessed on 5 May 2019).

60. Pan, X.; Saddler, J.N. Effect of replacing polyol by organosolv and kraft lignin on the property and structure of rigid polyurethane foam. *Biotechnol. Biofuels* **2013**, *6*, 12–21. [CrossRef]

61. Tavares, L.B.; Boas, C.V.; Schleder, G.R.; Nacas, A.M.; Rosa, D.S.; Santos, D.J. Bio-based polyurethane prepared from Kraft lignin and modified castor oil. *eXPRESS Pol. Lett.* **2016**, *10*, 927–940. [CrossRef]

62. Dohlen, S.; Braun, C.; Brodkorb, F.; Fischer, B.; Ilg, Y.; Kalbfleisch, K.; Kreyenschmidt, M.; Lorenz, R.; Kreyenschmidt, J. Effect of different packaging materials containing poly-[2-(tert-butylamino) methylstyrene] on the growth of spoilage and pathogenic bacteria on fresh meat. *Int. J. Food Microbiol.* **2017**, *257*, 91–100. [CrossRef]

63. Dohlen, S.; Braun, C.; Brodkorb, F.; Fischer, B.; Ilg, Y.; Kalbfleisch, K.; Kreyenschmidt, M.; Lorenz, R.; Robers, O.; Kreyenschmidt, J. Potential of the polymer poly-[2-(tert-butylamino) methylstyrene] as antimicrobial packaging material for meat products. *J. Appl. Microbiol.* **2016**, *4*, 1059–1070. [CrossRef]

64. Hüwe, C.; Schmeichel, J.; Brodkorb, F.; Dohlen, S.; Kalbfleisch, K.; Kreyenschmidt, M.; Lorenz, R.; Kreyenschmidt, J. Potential of antimicrobial treatment of linear low-density polyethylene with poly((tert-butyl-amino)-methyl-styrene) to reduce biofilm Formation in the Food industry. *Biofouling* **2018**, *34*, 378–387. [CrossRef]

65. Braun, C.; Dohlen, S.; Ilg, Y.; Brodkorb, F.; Fischer, B.; Heindirk, P.; Kalbfleisch, K.; Richter, T.; Robers, O.; Kreyenschmidt, M. Antimicrobial activity of intrinsic antimicrobial polymers based on poly((tertbutyl-amino)-methyl-styrene) against selected pathogenic and spoilage microorganisms relevant in meat processing facilities. *J. Antimicrob Agents* **2017**, *3*, 1000136. [CrossRef]

66. Song, W.; Ge, S. Application of antimicrobial nanoparticles in dentistry. *Molecules* **2019**, *24*, 1033. [CrossRef]

Article

Chitosan Cross-Linked Bio-based Antimicrobial Polypropylene Meshes for Hernia Repair Loaded with Levofloxacin HCl via Cold Oxygen Plasma

Noor Sanbhal [1,2], Yan Li [1,*], Awais Khatri [2] Mazhar Peerzada [2] and Lu Wang [1,*]

[1]   Key Laboratory of Textile Science and Technology of Ministry of Education, College of Textiles, Donghua University, 2999 North Renmin Road, Songjiang, Shanghai 201620, China; noor.sanbhal@faculty.muet.edu.pk

[2]   Department of Textile Engineering, Mehran University of Engineering and Technology, Jamshoro 76062, Sindh, Pakistan; awais.khatri@faculty.muet.edu.pk (A.K.); mazhar.peerzada@faculty.muet.edu.pk (M.P.)

*   Corresponding: yanli@dhu.edu.cn (Y.L.); wanglu@dhu.edu.cn (L.W.); Tel./Fax: +86-21-6779-2637 (L.W.)

Received: 18 December 2018; Accepted: 26 February 2019; Published: 4 March 2019

Abstract: Polypropylene (PP) large pore size nets have been most widely used implants for hernia repair. Nevertheless, the growth of bacteria within PP mesh pores after operation is a major reason of hernia recurrence. Secondly, pre-operative prophylaxis during mesh implantation has failed due to the hydrophobic nature of PP meshes. Herein, chitosan cross-linked and levofloxacin HCl incorporated, antimicrobial PP mesh devices were prepared using citric acid as a bio-based and green cross-linking agent. The inert PP mesh fibers were surface activated using O_2 plasma treatment at low pressure. Then, chitosan of different molecular weights (low and medium weight) were cross-linked with O_2 plasma activated surfaces using citric acid. Scanning electron microscopy (SEM), energy dispersive X-ray (EDX) spectroscopy, and Fourier transform infrared (FTIR) spectroscopy confirmed that chitosan was cross-linked with O_2 plasma-treated PP mesh surfaces and formed a thin layer of chitosan and levofloxacin HCl on the PP mesh surfaces. Moreover, antimicrobial properties of chitosan and levofloxacin HCl-coated PP meshes were investigated using an agar plate release method. The coated PP meshes demonstrated excellent antimicrobial inhibition zone up to 10 mm. Thus, modified PP meshes demonstrated sustained antimicrobial properties for six continuous days against *Staphylococcus aureus* (*SA*) and *Escherichia coli* (*EC*) bacteria.

Keywords: antimicrobial; polypropylene; chitosan; citric acid; cross-linked; cold plasma

---

1. Introduction

Light weight polypropylene (PP) mesh implantation for hernia repair has been performed to reinforce damaged tissues of the abdominal wall [1]. The recurrence rate of repaired hernia has been reduced marginally by using synthetic PP implants [2–4]. However, after PP mesh implantation, infection can be rare (1%–4%) [5,6], but subsequent failure of hernia mesh devices cannot be undervalued [7,8]. Reasons for infection may be the colonization of bacteria of the uneven knitting surfaces of PP meshes which may cause fistula formation around its pore size and result in the formation of granuloma [9]. In fact, pre-operative prophylaxis has no impact on mesh infection prevention because of non-absorbent PP characteristics [10,11]. Mesh infection is difficult to cure, considering continuous antibiotic therapy for a longer duration time, or it may result in the removal of implants [12]. Therefore, mesh infections should be cured in the very early stages, during mesh implantation [13].

Plasma treatment is an effective method used to modify the surfaces of biomaterials [14–16]. Among other plasma processes, cold oxygen plasma has been reported as a suitable process [17–19] to modify PP fiber surfaces without changing bulk properties [20,21]. However, different surface coatings for PP meshes have been reported [15,22–26], but few authors have suggested green bio-based drug carriers such as chitosan for prolonged antimicrobial effects [27–29]. Chitosan is obtained from chitin, a very cheap product commonly available as shellfish waste. Chitosan is a natural polymer which has great importance as a biomaterials polymer due to its exceptional biological properties, such as being a non-toxic biodegradable and biocompatible drug carrier. Chitosan is commonly used in biomedical devices because of its excellent antimicrobial properties against gram-positive and gram-negative bacteria. Moreover, it is a carbohydrate used as an advanced active species to coordinate with transit metals [29]. Chitosan can carry broad spectrum soluble antibiotic drugs and release them after a suitable duration of time for mesh infection prevention. Thus, the polymer having antibiotic properties itself and being used as an antibiotic drug carrier may be of great interest in the application of hernia mesh surface functional materials [27]. The existence of hydroxyl and amino groups on chitosan gives it the advantage of being able to cross-link with a number of chemicals to form amide and ester bonds [30–32]. Moreover, levofloxacin HCl is a soluble drug commonly used for infection prevention with suitable release properties and it is capable of inhibiting both types (gram-negative and gram-positive) of bacteria [33].

Cakmak et al. have reported satisfactory results of chitosan- and triclosan-coated PP meshes [34]. Nevertheless, growing concern around triclosan limits its application [35]. Moreover, Avetta et al. have reported surface functionalisation of PP meshes with chitosan and ciprofloxacin but with antibacterial properties lasting only four days [27].

However, chitosan can be cross-linked using different cross-linking agents such as formaldehyde and glutaraldehyde. Nevertheless, these chemicals are toxic and the biocompatibility of the yield product is the primary and most basic requirement for all medical devices. Citric acid has been used for biomedical applications [36–38] and it has been reported that citric acid is a safe and bio-based green crosslinking agent for chitosan polymers [39].

In our previous work, we treated PP meshes with cold oxygen plasma, after which two steps grafting (with hexamethylene diisocyanate and cyclodextrin) were performed for β-cyclodextrin incorporation; afterwards, levofloxacin HCl was loaded into the β-cyclodextrins (CD) cavity [40]. Herein, a simple textile based one bath padding method was selected to incorporate a chitosan and levofloxacin HCl coating onto oxygen plasma-treated meshes using citric acid.

Polypropylene (PP) meshes were first surface activated using O_2 plasma treatment at low pressure, after which the PP meshes were padded (pick up 90%) with a prepared solution of chitosan and levofloxacin HCl containing citric acid as a cross-linking agent, as shown in Figure 1. The padded PP mesh devices were dried (40 °C) and cured (70 °C) in an oven for 10 min. The surface morphology, chemical composition, and structural changes of the plasma treated chitosan and levofloxacin HCl-coated PP meshes were characterized using scanning electron microscopy (SEM), energy dispersive X-ray (EDX) spectroscopy, and Fourier transform infrared (FTIR) spectroscopy. Results revealed that the PP mesh was connected with chitosan through the oxygen plasma treatment. Thus, a thin layer of chitosan and levofloxacin HCl was observed on the surfaces of the PP meshes. Furthermore, antimicrobial properties of chitosan and levofloxacin HCl-coated PP meshes were performed via an agar diffusion plate by release properties. Thus, the chitosan and levofloxacin-modified devices demonstrated an excellent antimicrobial zone of inhibition and sustained antimicrobial release properties for six continuous days.

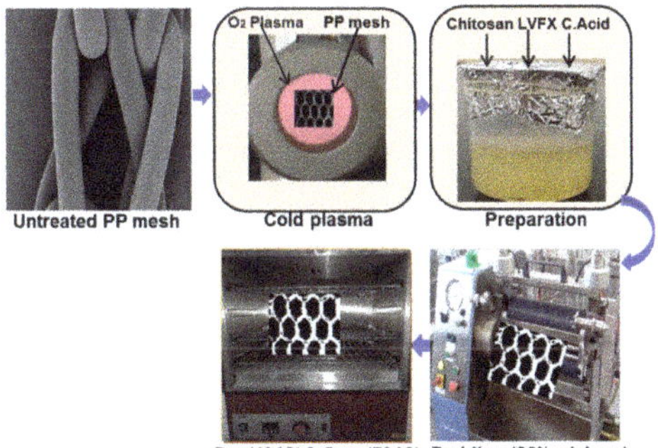

Figure 1. Illustrations and experimental design of chitosan and levofloxacin HCl-coated polypropylene (PP) meshes.

2. Materials and Methods

2.1. Materials

Light weight (27 g/m^2) PP mono filament meshes with pore sizes 2.5 mm $\times$ 3.5 mm were purchased from Nantong Newtec Textile and Chemical Fiber Co. Ltd. China (Nantong, China). Levofloxacin hydrochloride (98%) was purchased from Energy Chemicals Shanghai China (Shanghai, China). Citric acid monohydrate and acetic acid were received of analytical reagents.

2.2. Surface Functionalization of PP with Oxygen Plasma

PP mesh devices (10 cm $\times$ 10 cm) were functionalized using cold oxygen plasma (HD-300) at low pressure. All samples were surface activated for 5 min at 45 W. The oxygen gas flow was kept at less than 0.2 bar.

2.3. Preparation of Chitosan and Levofloxacin HCl Solution and Coating onto PP Meshes

The antibacterial solution was prepared following the pad dry method described by Avetta et al. [27]. Chitosan of low molecular weight (CH-LMW) and medium molecular weight (CH-MMW) were dissolved in deionized water with acetic acid. The solution was stirred for 12 h at room temperature using a magnetic stirrer at 150 rpm. Then, levofloxacin HCl was poured into the solution of chitosan and stirred for a further 2 h to mix the antimicrobial homogeneously. The overall concentration maintained in the solution was chitosan 2.0 wt % (low and medium), levofloxacin HCl (0.2%), citric acid (2%) and acetic acid (0.5%) was added to dissolve the chitosan.

Plasma-treated meshes were dipped in the chitosan solution and padded through horizontal padding at 90% pick up (1.2 bar pressure). The mesh devices were padded twice to get an even coating on the whole mesh surfaces. Moreover, coated samples were first dried at 40 °C and then cured at 70 °C for 10 min to perform the cross linking of chitosan with the plasma treated fibers.

3. Characterization

3.1. SEM and EDX

Chitosan and levofloxacin HCl-modified PP meshes were evaluated for surface morphology by scanning electron microscopy (Quanta SEM 250, FEI, Waltham, MA, USA). Samples were used for coating platinum (Pt) before SEM scanning. Moreover, element analysis of chitosan and levofloxacin HCl-modified PP mesh devices was performed using energy dispersive X-ray spectroscopy (ISIS 300, Oxford Instruments, Oxfordshire, UK).

3.2. FTIR

The surface structures of the modified and control meshes were investigated using Fourier transform infrared spectroscopy (Nicolet 6700, Thermo Fisher Scientific, Waltham, MA, USA) of attenuated total reflection (ATR). The FTIR range used to analysis the structure was between 4000–500 cm^{-1}.

3.3. Differential Scanning Calorimetry (DSC) and X-ray Diffraction (XRD)

A DSC (Pyris, Perkin Elemer, Waltham, MA, USA) test was performed to get the melting temperature of the chitosan and levofloxacin HCl-coated and untreated samples. All samples were scanned in the temperature range of 30–300 °C at 20 °C/min.

Moreover, chitosan and levofloxacin-coated and untreated samples were characterized using an X-ray diffractometer (Rigaku D/MAX 2550/PC, Tokyo, Japan). The range of crystallization analysis was 5° to 60° (2θ) and the testing rate was set at 0.02°/min.

3.4. Antibacterial Activity

The antibacterial activity of chitosan (low and medium molecular weight) was investigated using a simple agar diffusion plate test method. A specific amount of 400 μL of bacteria (*Staphylococcus aureus* (*SA*) and *Escherichia coli* (*EC*)) of 1×10^8 colony forming units (CFU)/mL was poured on agar plates. Then, treated and untreated samples (1 cm × 1 cm) were placed onto the center of the agar plates and all samples were incubated in an oven at 37 °C for 24 h [41]. The zone of inhibition of each sample was measured in all four directions and described as an average antibacterial inhibition zone value. The formula for the zone of inhibition was $C = (K - B)/2$ where C = inhibition zone, K = inhibition zone after incubation (24 h), and B is the original sample (1 × 1) without antibacterial activity.

Furthermore, PP-untreated and chitosan-treated samples were analyzed for their antibacterial release properties. Each day samples were transferred to new agar plates and fresh bacteria were poured. Thus, after 24 h the inhibition zone was measured and compared with the previous one. The antibacterial release performance was continued until the modified meshes sustained antibacterial activity.

3.5. Statistical Analysis

The standard deviation and mean are reported in Figure 8. However, standard bars in the figures represent standard deviation. One-way single factor ANOVA was performed to find out the actual differences for each sample. The figure data is marked with $p < 0.001$ (***), $p < 0.01$ (**), and $p < 0.05$ (*). Thus, the value of $p < 0.05$ (*) was chosen as a confidence interval value.

4. Results and Discussion

4.1. Chitosan and Levofloxacin Coating onto PP Mesh Surfaces

Chitosan was cross-linked with PP meshes using citric acid monohydrate, as shown in the reaction scheme (Figure 2). Chitosan of low molecular weight and chitosan of medium molecular weight with levofloxacin HCl were prepared using acetic acid and stirred for 12 h, after which citric acid monohydrate was poured into the solutions prior to 2 h of coating. The coating solutions containing CH-LMW and CH-MMW were separately padded onto O$_2$ plasma activated surfaces of PP meshes. The coated meshes were dried and cured at 70 °C for 10 min to cross-link chitosan with the PP mesh surfaces. The average corresponding weight of the samples of low molecular weight (2.05 ± 1.3%) and medium molecular weight (4.1% ± 0.8%) was increased. The result was that a thin layer of chitosan with levofloxacin HCl was obtained onto the plasma activated PP fiber surfaces. Thus, plasma-treated PP meshes of CH-MMW received more amounts of surface coating in comparison to CH-MMW.

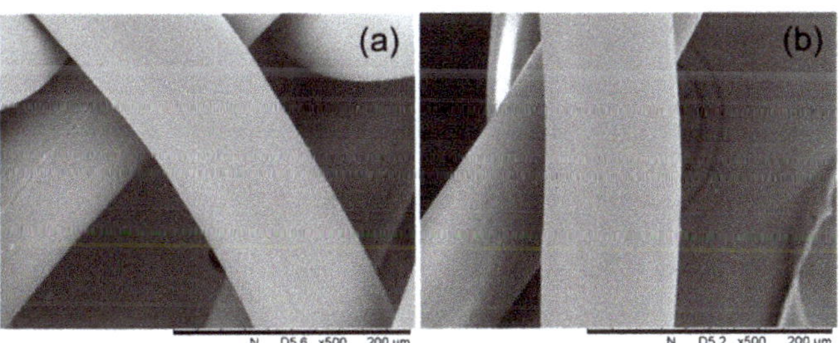

Figure 2. Reaction scheme of cross-linked chitosan with oxygen plasma-treated PP meshes using citric acid.

4.2. Surface Morphology of Chitosan and Levofloxacin HCl-Modified PP Meshes

SEM images of chitosan and levofloxacin HCl-modified plasma-treated and -untreated PP meshes are shown in Figure 3. It can be observed that the untreated PP meshes display noticeable line marks on their surfaces (Figure 3a) with bright surface structures, but after oxygen plasma treatment such line marks are missing, showing (Figure 3b) relatively dull and even surfaces. The plasma-treated fibers have regular surfaces in comparison to the untreated meshes. Moreover, the low molecular weight chitosan and levofloxacin-coated meshes display (Figure 3c) an even thin layer across the whole spherical diameter of the fibers. However, it can be observed that the medium weight chitosan and levofloxacin HCl completely coated the (Figure 3d) plasma-activated PP fibers with a thick and sticky layer. It can be seen that the sticky layer (the CH-MMW) stretches across the whole surface of the PP fibers, showing a more even coating than the CH-LMW coating. Thus, it can be summarized that medium molecular weight chitosan can coat PP fibers more effectively. The reason for the thick and even surface coating may be the oxygen plasma treatment, which may enhance the adhesion of the chitosan and levofloxacin HCl coating [19].

Figure 3. *Cont.*

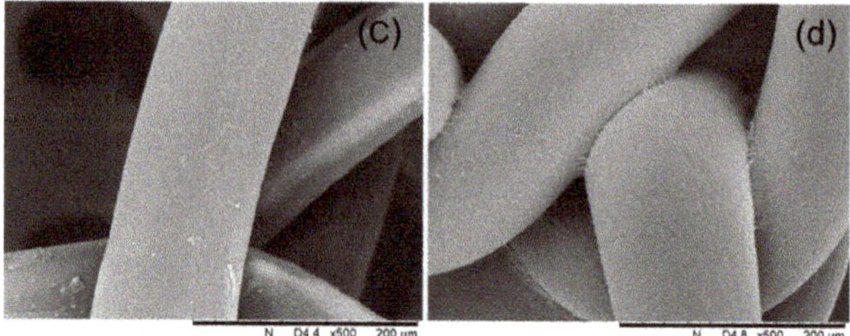

Figure 3. Scanning electron microscopy (SEM) images: (**a**) PP-untreated, (**b**) oxygen plasma-treated, (**c**) chitosan of low molecular weight (CH-LMW)-coated PP meshes, and (**d**) chitosan of medium molecular weight (CH-MMW) and levofloxacin HCl-coated PP meshes. All samples were scanned at a magnification of 500×.

4.3. Characterization of Chitosan and Levofloxacin HCl-Modified PP Devices

Figure 4 displays energy dispersive X-ray spectroscopy peaks for the identification of the surface chemical structures of the PP meshes and the plasma-treated and chitosan and levofloxacin HCl-modified mesh devices. The PP control meshes displayed a 100% carbon peak at 0.4 keV. Nevertheless, the plasma-treated meshes confirmed a carbon atom at a similar 0.4 keV mark but an additional oxygen (O) atom peak (2.9%) was observed at 0.7 keV. Moreover, CH-LMW-coated PP meshes displayed an increase in oxygen (13.31%) atoms and additionally, two peaks of nitrogen and fluorine can be observed around 0.8 keV. Furthermore, medium molecular weight chitosan and levofloxacin HCl-coated samples displayed same atoms similar to CH-LMW but with increased atomic weight percentages, indicating the good efficiency of the chitosan layer when making it onto the PP surfaces. These results are in accord with published paper [27], except that we received an additional peak of levofloxacin HCl, which is the most commonly used antimicrobial for infection prevention. Thus, cold oxygen plasma is shown to be an important process regarding enhancing surface adhesion for the coating of chitosan and levofloxacin HCl onto PP mesh surfaces.

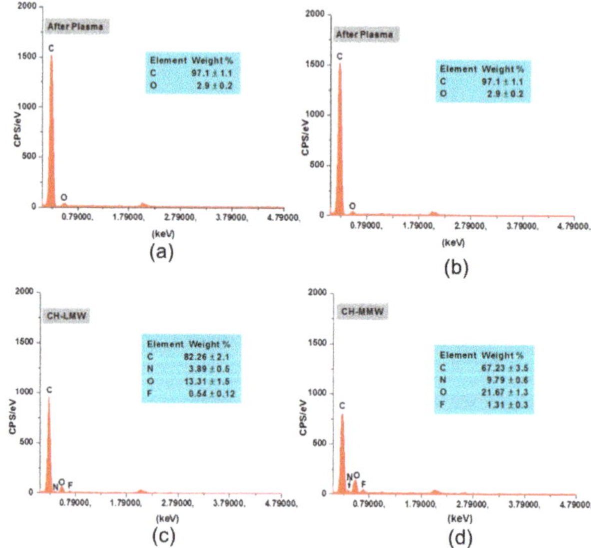

Figure 4. Energy dispersive X-ray (EDX) spectra: (**a**) PP control, (**b**) oxygen plasma-functionalized, (**c**) low molecular weight chitosan and levofloxacin, and (**d**) medium molecular weight chitosan and levofloxacin HCl-coated PP meshes.

FTIR spectra of chitosan and levofloxacin HCl-coated meshes are shown in Figure 5. The PP control fibers displayed identical peaks at 2951, 2918, 1453, and 1378 cm^{-1} [42,43]. However, when excluding the PP fiber original peaks, a new peak (Figure 5a) at 3347 cm^{-1} may be observed, which may be due to the oxygen plasma treatment which provides an OH group to the PP fiber surfaces. Moreover, as shown in Figure 5b, chitosan-coated modified meshes show a slight change in the vibration band at 3398 cm^{-1}, but the peak height was more in alignment with the oxygen plasma-treated fibers. The chitosan-coated modified PP meshes also show an additional cross-linking peak at 1715 cm^{-1}, which may be due to the formation of a carbonyl group. However, amide I and amide II were observed at 1625 and 1215 cm^{-1}, respectively. The C–O stretch of chitosan was seen within the fingerprint region at 1070 cm^{-1} for the medium-weight chitosan-coated meshes. These results are in accord with published paper [27], but most notably in our work we found extra peaks at 1715 cm^{-1}, which are due to the formation of a carbonyl group. The reason for the formation of a peak at 1070 cm^{-1} specially for the CH-MMW meshes may be due to the better coating efficiency of chitosan onto the oxygen-treated PP meshes in comparison with the CH-LMW coating.

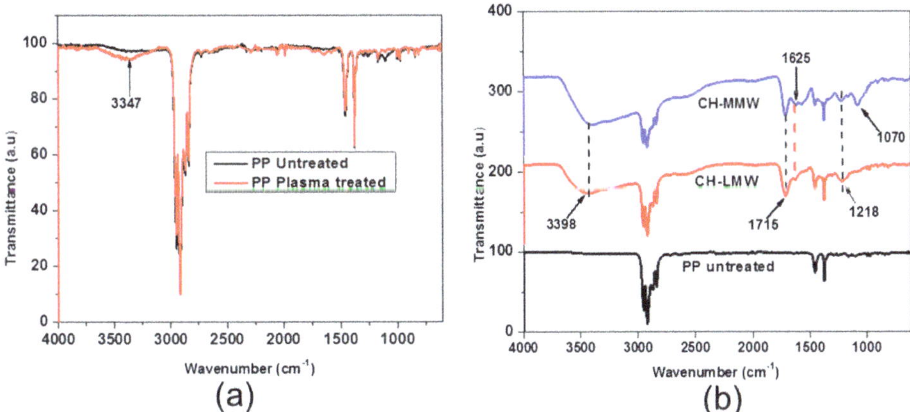

Figure 5. Fourier transform infrared (FTIR) spectra (attenuated total reflection (ATR)): (**a**) PP control and oxygen plasma-treated PP meshes; (**b**) PP control, CH-LMW, and CH-MMW and levofloxacin-coated PP meshes.

4.4. Thermal and Structural Properties

As shown in Figure 6a, the PP meshes without treatment, and those coated with chitosan and levofloxacin HCl, have no identical differences except that the control PP mesh has more peak height. Moreover, all three samples—PP control, CH-LMW, and CH-MMW—have almost the same melting temperatures, these being 148.5, 148.9, and 149.05 °C, respectively. Thus, there is a slight increase in melting temperature as chitosan coating is applied to the PP mesh surfaces. Overall, there is no difference in melting temperature between the treated and untreated PP meshes.

Figure 6b displays XRD patterns of chitosan-coated and non-coated PP meshes. It can be observed that the PP meshes coated with low molecular weight chitosan, medium molecular weight chitosan, and those which are untreated exhibited similar pattern peaks and crystal structures for the 2θ range 5°–60°. Therefore, there are no identical structural changes before and after the surface modification of the PP meshes.

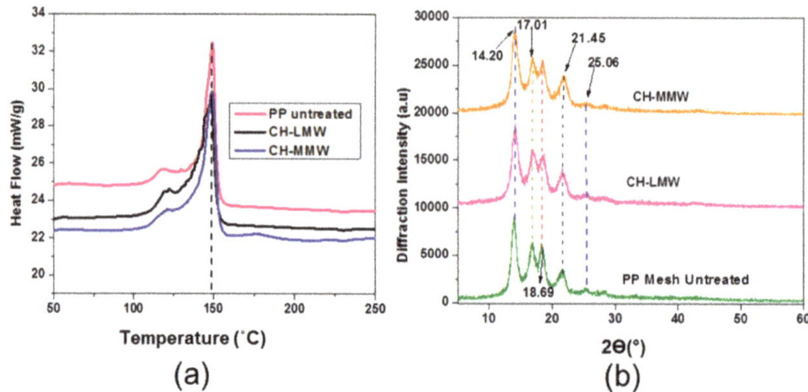

Figure 6. Thermal and structural properties: (**a**) differential scanning calorimetry (DSC) analysis and (**b**) X-ray diffraction (XRD) patterns of treated and untreated PP meshes.

4.5. Antibacterial Activity

Antibacterial properties of the chitosan and levofloxacin HCl (CH-LMW and CH-MMW) and untreated meshes are shown in Figure 7. Untreated PP mesh fibers were unable to resist bacterial (*EC* and *SA*) growth. Thus, there is no inhibition zone (Figure 7a,b) around the PP control samples. Nevertheless, the CH-LMW and levofloxacin HCl-coated PP meshes demonstrated suitable antibacterial properties. The CH-LMW-coated meshes displayed an average inhibition zone of 8.1 and 8.6 mm for *SA* and *EC*, respectively. However, in the case of the CH-MMW and levofloxacin coated samples, these demonstrated better average inhibition zones of 10.1 and 10.9 mm for *SA* and *EC*, respectively. Thus, the CH-MMW samples demonstrated a bigger average inhibition zone than the CH-LMW samples.

Thus, it was confirmed that PP control does not exhibit antibacterial properties. For this reason, the CH-LMW and CH-MMW samples were assessed for their antimicrobial release properties. Both samples (CH-LMW and CH-MMW) were further tested (Figure 8) against *SA* and *EC* by a release method. As shown in Figure 8a, the average initial inhibition zones for the CH-LMW samples for the first day against *SA* and *EC* were 8.01 and 8.533 mm, respectively. CH-LMW demonstrated antibacterial release properties for four days against *SA* and *EC* bacteria. *EC* displayed a bigger inhibition zone than *SA* over the four days. However, the minimum inhibition zones for *SA* (0.5 mm) and *EC* (0.8 mm) were measured on the fourth day.

Moreover, antimicrobial properties of CH-MMW samples sustained (Figure 8b) for 6 days. CH-MMW samples displayed 11 and 10.1 mm inhibition zone for *SA* and *EC*, respectively. This was greater inhibition zone than CH-LMW. However, similarly like CH-LMW, EC shown bigger inhibition zone during entire 6 days but on 4th day zone of inhibition was almost same for *SA* and *EC*. Minimum average inhibition zone for SA (0.3 mm) and EC (1.2 mm) were seen on day 6. According to previous published papers author reported chitosan functionalized PP meshes [29] only 4 days of drug release was achieved [27]. However, antimicrobial results for CH-MMW shown better results and sustained antibacterial release up to 6 days. This may be due to the surface functionalization of PP meshes with oxygen plasma as literature described that cold oxygen plasma is a more suitable surface treatment than other plasma treatments [19,40]. Secondly citric acid is a good cross-linker and has been used commonly to connect hydroxyl group with chitosan [30]. Therefore, formation of hydroxyl group on PP surfaces given advantage to connect oxygen plasma treated PP fibers with chitosan. The results of SEM, FTIR and antimicrobial release method proved that chitosan and levofloxacin layer was cross-linked with plasma activated surfaces which gave better results of antimicrobial.

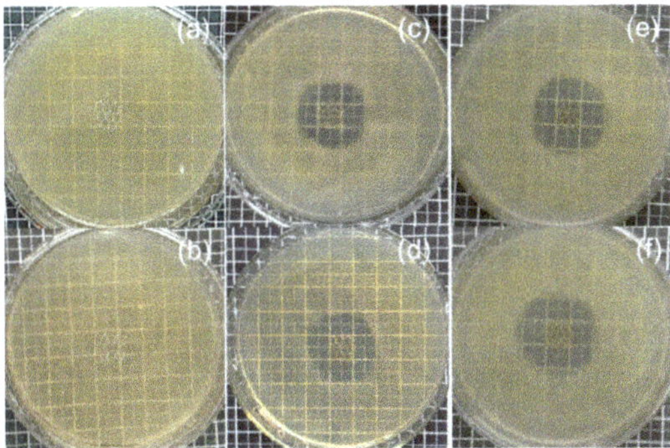

Figure 7. Antibacterial activity by inhibition zone: (**a,b**) PP control, (**c,d**) low molecular weight chitosan and levofloxacin-coated, (**e,f**) medium molecular weight chitosan and levofloxacin HCl-coated. Note: the top row is *SA* and the bottom row is *EC*.

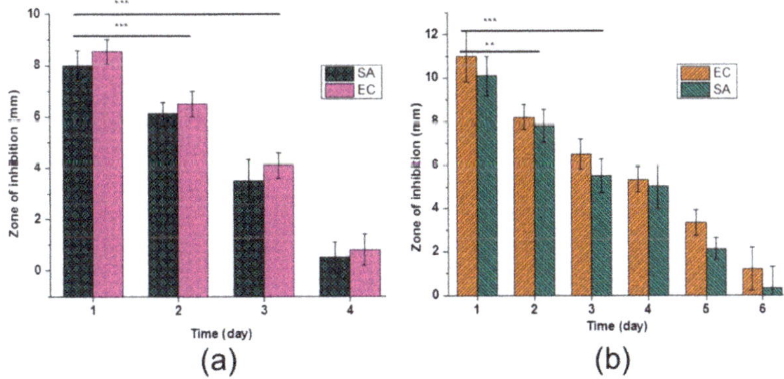

Figure 8. Antibacterial activity by release method (**a**) low molecular weight chitosan and levofloxacin coated CH-LMW (**b**) Medium molecular weight and levofloxacin HCL coated PP meshes (CH-MMW).

5. Conclusions

In this study, chitosan of low molecular weight and chitosan of medium molecular weight with levofloxacin HCl were successfully coated onto oxygen plasma-treated PP mesh fiber surfaces. Plasma activation created adhesion on the surfaces of the PP fibers, which was utilized to connect PP meshes with chitosan in presence of citric acid. The result was a thin layer of chitosan and levofloxacin HCl coating the PP meshes. FTIR confirmed that chitosan was successfully attached and cross-linked with the PP mesh fibers.

Moreover, the chitosan and levofloxacin HCl-modified PP meshes demonstrated excellent antimicrobial inhibition zones and antimicrobial release properties which were sustained for six days. Thus, a green and bio-based chitosan with suitable antimicrobial properties could be used for mesh infection prevention during hernia repair.

Author Contributions: Conceptualization, N.S. and Y.I.; Methodology, N.S.; Software, N.S.; Validation, A.K. and N.S.; Writing—Original Draft Preparation, M.P. and N.S.; Writing-Review and Editing, L.W.

Funding: This work was supported by the 111 project "Biomedical Textile Material (No. B07024) Science and Technology", the National Key Research Development Program (No. 2016YFB 0303300-03) of China, and the Fundamental Research (Nos. 17D110111 and 2232018G-01) Funds for the Central Universities.

Acknowledgments: Thanks to Rafique Ahmed lakho for his help during the experiments.

Conflicts of Interest: The authors declare no conflict of interest.

References

1. Miao, L.; Wang, F.; Wang, L.; Zou, T.; Brochu, G.; Guidoin, R. Physical characteristics of medical textile prostheses designed for hernia repair: A comprehensive analysis of select commercial devices. *Materials* **2015**, *8*, 8148–8168. [CrossRef] [PubMed]
2. Greca, F.; Paula, J.; Biondo-Simões, M.; Costa, F.; Silva, A.; Time, S.; Mansur, A. The influence of differing pore sizes on the biocompatibility of two polypropylene meshes in the repair of abdominal defects. *Hernia* **2001**, *5*, 59–64. [PubMed]
3. Hazebroek, E.J.; Ng, A.; Yong, D.H.; Berry, H.; Leibman, S.; Smith, G.S. Evaluation of lightweight titanium-coated polypropylene mesh (TiMesh) for laparoscopic repair of large hiatal hernias. *Surg. Endosc.* **2008**, *22*, 2428–2432. [CrossRef] [PubMed]
4. Jerabek, J.; Novotny, T.; Vesely, K.; Cagas, J.; Jedlicka, V.; Vlcek, P.; Capov, I. Evaluation of three purely polypropylene meshes of different pore sizes in an onlay position in a New Zealand white rabbit model. *Hernia* **2014**, *18*, 855–864. [CrossRef] [PubMed]
5. Perez-Kohler, B.; Bayon, Y.; Bellon, J.M. Mesh infection and hernia repair: A review. *Surg. Infect.* **2016**, *17*, 124–137. [CrossRef] [PubMed]
6. Guillaume, O.; Perez-Tanoira, R.; Fortelny, R.; Redl, H.; Moriarty, T.F.; Richards, R.G.; Eglin, D.; Petter Puchner, A. Infections associated with mesh repairs of abdominal wall hernias: Are antimicrobial biomaterials the longed-for solution? *Biomaterials* **2018**, *167*, 15–31. [CrossRef] [PubMed]
7. Kulaga, E.; Ploux, L.; Balan, L.; Schrodj, G.; Roucoules, V. Mechanically responsive antibacterial plasma polymer coatings for textile biomaterials. *Plasma Process. Polym.* **2014**, *11*, 63–79. [CrossRef]
8. Knetsch, M.L.; Koole, L.H. New strategies in the development of antimicrobial coatings: The example of increasing usage of silver and silver nanoparticles. *Polymers* **2011**, *3*, 340–366. [CrossRef]
9. Sanbhal, N.; Miao, L.; Xu, R.; Khatri, A.; Wang, L. Physical structure and mechanical properties of knitted hernia mesh materials: A review. *J. Ind. Text.* **2018**, *48*, 333–360. [CrossRef]
10. Mazaki, T.; Mado, K.; Masuda, H.; Shiono, M.; Tochikura, N.; Kaburagi, M. A randomized trial of antibiotic prophylaxis for the prevention of surgical site infection after open mesh-plug hernia repair. *Am. J. Surg.* **2014**, *207*, 476–484. [CrossRef] [PubMed]
11. Mazaki, T.; Mado, K.; Masuda, H.; Shiono, M. Antibiotic prophylaxis for the prevention of surgical site infection after tension-free hernia repair: A Bayesian and frequentist meta-analysis. *J. Am. Coll. Surg.* **2013**, *217*, 788–801. [CrossRef] [PubMed]
12. Zhang, Z.; Tang, J.; Wang, H.; Xia, Q.; Xu, S.; Han, C.C. Controlled antibiotics release system through simple blended electrospun fibers for sustained antibacterial effects. *ACS Appl. Mater. Interfaces* **2015**, *7*, 26400–26404. [CrossRef] [PubMed]
13. Guillaume, O.; Garric, X.; Lavigne, J.P.; Van Den Berghe, H.; Coudane, J. Multilayer, degradable coating as a carrier for the sustained release of antibiotics: Preparation and antimicrobial efficacy in vitro. *J. Control. Release* **2012**, *162*, 492–501. [CrossRef] [PubMed]
14. Khelifa, F.; Ershov, S.; Habibi, Y.; Snyders, R.; Dubois, P. Free-radical-induced grafting from plasma polymer surfaces. *Chem. Rev.* **2016**, *116*, 3975–4005. [CrossRef] [PubMed]
15. Nisticò, R.; Rosellini, A.; Rivolo, P.; Faga, M.G.; Lamberti, R.; Martorana, S.; Castellino, M.; Virga, A.; Mandracci, P.; Malandrino, M.; et al. Surface functionalisation of polypropylene hernia-repair meshes by RF-activated plasma polymerisation of acrylic acid and silver nanoparticles. *Appl. Surf. Sci.* **2015**, *328*, 287–295. [CrossRef]
16. Nisticò, R.; Magnacca, G.; Faga, M.G.; Gautier, G.; D'Angelo, D.; Ciancio, E.; Lamberti, R.; Martorana, S. Effect of atmospheric oxidative plasma treatments on polypropylenic fibers surface: Characterization and reaction mechanisms. *Appl. Surf. Sci.* **2013**, *279*, 285–292. [CrossRef]
17. Jelil, R.A. A review of low-temperature plasma treatment of textile materials. *J. Mater. Sci.* **2015**, *50*, 5913–5943. [CrossRef]
18. Lai, J.; Sunderland, B.; Xue, J.; Yan, S.; Zhao, W.; Folkard, M.; Michael, B.D.; Wang, Y. Study on hydrophilicity of polymer surfaces improved by plasma treatment. *Appl. Surf. Sci.* **2006**, *252*, 3375–3379. [CrossRef]
19. Sorrentino, L.; Carrino, L.; Napolitano, G. Oxygen cold plasma treatment on polypropylene: Influence of process parameters on surface wettability. *Surf. Eng.* **2013**, *23*, 247–252. [CrossRef]
20. Shishoo, R. *Plasma Technologies for Textiles*; Woodhead publishing limited: Sawston, UK, 2007.

21. Fauland, G.; Constantin, F.; Gaffar, H.; Bechtold, T. Production scale plasma modification of polypropylene baselayer for improved water management properties. *J. Appl. Polym. Sci.* **2015**, *132*, 41294. [CrossRef]

22. Perez-Kohler, B.; Fernandez-Gutierrez, M.; Pascual, G.; Garcia-Moreno, F.; San Roman, J.; Bellon, J.M. In vitro assessment of an antibacterial quaternary ammonium-based polymer loaded with chlorhexidine for the coating of polypropylene prosthetic meshes. *Hernia* **2016**, *20*, 869–878. [CrossRef] [PubMed]

23. Labay, C.; Canal, J.M.; Modic, M.; Cvelbar, U.; Quiles, M.; Armengol, M.; Arbos, M.A. Antibiotic-loaded polypropylene surgical meshes with suitable biological behaviour by plasma functionalization and polymerization. *Biomaterials* **2015**, *71*, 132–144. [CrossRef] [PubMed]

24. Harth, K.C.; Rosen, M.J.; Thatiparti, T.R.; Jacobs, M.R.; Halaweish, I.; Bajaksouzian, S.; Furlan, J.; von Recum, H.A. Antibiotic-releasing mesh coating to reduce prosthetic sepsis: An in vivo study. *J. Surg. Res.* **2010**, *163*, 337–343. [CrossRef] [PubMed]

25. Gorgieva, S.; Modic, M.; Dovgan, B.; Kaisersberger-Vincek, M.; Kokol, V. Plasma-activated polypropylene mesh-gelatin scaffold composite as potential implant for bioactive hernia treatment. *Plasma Process. Polym.* **2015**, *12*, 237–251. [CrossRef]

26. Muzio, G.; Miola, M.; Perero, S.; Oraldi, M.; Maggiora, M.; Ferraris, S.; Vernè, E.; Festa, V.; Festa, F.; Canuto, R.A.; et al. Polypropylene prostheses coated with silver nanoclusters/silica coating obtained by sputtering: Biocompatibility and antibacterial properties. *Surf. Coat. Technol.* **2017**, *319*, 326–334. [CrossRef]

27. Avetta, P.; Nisticò, R.; Faga, M.G.; D'Angelo, D.; Boot, E.A.; Lamberti, R.; Martorana, S.; Calza, P.; Fabbri, D.; Magnacca, G. Hernia-repair prosthetic devices functionalised with chitosan and ciprofloxacin coating: Controlled release and antibacterial activity. *J. Mater. Chem. B* **2014**, *2*, 5287–5294. [CrossRef]

28. Udpa, N.; Iyer, S.R.; Rajoria, R.; Breyer, K.E.; Valentine, H.; Singh, B.; McDonough, S.P.; Brown, B.N.; Bonassar, L.J.; Gao, Y. Effects of chitosan coatings on polypropylene mesh for implantation in a rat abdominal wall model. *Tissue Eng. Part A* **2013**, *19*, 2713–2723. [CrossRef] [PubMed]

29. Nisticò, R.; Faga, M.G.; Gautier, G.; Magnacca, G.; D'Angelo, D.; Ciancio, E.; Piacenza, G.; Lamberti, R.; Martorana, S. Physico-chemical characterization of functionalized polypropylenic fibers for prosthetic applications. *Appl. Surf. Sci.* **2012**, *258*, 7889–7896. [CrossRef]

30. Aubert-Viard, F.; Martin, A.; Chai, F.; Neut, C.; Tabary, N.; Martel, B.; Blanchemain, N. Chitosan finishing nonwoven textiles loaded with silver and iodide for antibacterial wound dressing applications. *Biomed. Mater.* **2015**, *10*, 015023. [CrossRef] [PubMed]

31. Shweta, A.; Sonia, P. Pharmaceutical relevance of cross-linked chitosan in microparticulate drug delivery. *Int. Res. J. Pharm.* **2013**, *4*, 45–51.

32. Abraham, S.; Rajamanickam, D.; Srinivasan, B. Preparation, characterization and cross-linking of chitosan by microwave assisted synthesis. *Sci. Int.* **2018**, *6*, 18–30. [CrossRef]

33. Ren, Q.; Zhu, X. Methyl-beta-cyclodextrin/cetyltrimethyl ammonium bromide synergistic sensitized fluorescence method for the determination of levofloxacin. *J. Fluoresc.* **2016**, *26*, 671–677. [CrossRef] [PubMed]

34. Cakmak, A.; Cirpanli, Y.; Bilensoy, E.; Yorganci, K.; Calis, S.; Saribas, Z.; Kaynaroglu, V. Antibacterial activity of triclosan chitosan coated graft on hernia graft infection model. *Int. J. Pharm.* **2009**, *381*, 214–219. [CrossRef] [PubMed]

35. Sanbhal, N.; Mao, Y.; Sun, G.; Xu, R.F.; Zhang, Q.; Wang, L. Surface modification of polypropylene mesh devices with cyclodextrin via cold plasma for hernia repair: Characterization and antibacterial properties. *Appl. Surf. Sci.* **2018**, *439*, 749–759. [CrossRef]

36. Blanchemain, N.; Haulon, S.; Martel, B.; Traisnel, M.; Morcellet, M.; Hildebrand, H.F. Vascular PET prostheses surface modification with cyclodextrin coating: Development of a new drug delivery system. *Eur. J. Vasc. Endovasc. Surg.* **2005**, *29*, 628–632. [CrossRef] [PubMed]

37. Laurent, T.; Kacem, I.; Blanchemain, N.; Cazaux, F.; Neut, C.; Hildebrand, H.F.; Martel, B. Cyclodextrin and maltodextrin finishing of a polypropylene abdominal wall implant for the prolonged delivery of ciprofloxacin. *Acta Biomater.* **2011**, *7*, 3141–3149. [CrossRef] [PubMed]

38. Debbabi, F.; Gargoubi, S.; Hadj Ayed, M.A.; Abdessalem, S.B. Development and characterization of antibacterial braided polyamide suture coated with chitosan-citric acid biopolymer. *J. Biomater. Appl.* **2017**, *32*, 384–398. [CrossRef] [PubMed]

39. Varshosaz, J.; Alinagari, R. Effect of citric acid as cross-linking agent on insulin loaded chitosan microspheres. *Iran. Polym. J.* **2005**, *14*, 647–656.

40. Sanbhal, N.; Saitaer, X.; Li, Y.; Mao, Y.; Zou, T.; Sun, G.; Wang, L. Controlled levofloxacin release and antibacterial properties of β-cyclodextrins-grafted polypropylene mesh devices for hernia repair. *Polymers* **2018**, *10*, 493. [CrossRef]
41. Sanbhal, N.; Mao, Y.; Sun, G.; Li, Y.; Peerzada, M.; Wang, L. Preparation and characterization of antibacterial polypropylene meshes with covalently incorporated β-cyclodextrins and captured antimicrobial agent for hernia repair. *Polymers* **2018**, *10*, 58. [CrossRef]
42. Nava-Ortiz, C.A.; Alvarez-Lorenzo, C.; Bucio, E.; Concheiro, A.; Burillo, G. Cyclodextrin-functionalized polyethylene and polypropylene as biocompatible materials for diclofenac delivery. *Int. J. Pharm.* **2009**, *382*, 183–191. [CrossRef] [PubMed]
43. Sarau, G.; Bochmann, A.; Lewandowska, R.; Christianse, S. From micro– to macro–Raman spectroscopy: Solar silicon for a case study. In *Advanced Aspects of Spectroscopy*; Farrukh, M.A., Ed.; IntechOpen: London, UK, 2012; pp. 221–246.

Article

Polydopamine-Inspired Surface Modification of Polypropylene Hernia Mesh Devices via Cold Oxygen Plasma: Antibacterial and Drug Release Properties

Xiakeer Saitaer [1,2], Noor Sanbhal [2,3], Yansha Qiao [2], Yan Li [2,*], Jing Gao [2], Gaetan Brochu [4], Robert Guidoin [2,4], Awais Khatri [3] and Lu Wang [2,*]

1   College of Textiles and Fashion, Xinjiang University, 666 Sheng Li Road, Tian Shan, Wulumuqi 830046, China; xaker2@163.com
2   Key Laboratory of Textile Science and Technology of Ministry of Education, Key Laboratory of Biomedical Textile Materials and Technology in Textile Industry, College of Textiles, Donghua University, 2999 North Renmin Road, Songjiang, Shanghai 201620, China; noor.sanbhal@faculty.muet.edu.pk (N.S.); qiaoys233@163.com (Y.Q.); gao2001jing@dhu.edu.cn (J.G.); robert.guidoin@fmed.ulaval.ca (R.G.)
3   Department of Textile Engineering, Mehran University of Engineering and Technology, Jamshoro, Sindh 76062, Pakistan; awais.khatri@faculty.muet.edu.pk
4   Department of Surgery, University of Laval, Quebec, QC G1V 0A6, Canada; gaetan.brochu@chg.ulaval.ca
*   Correspondence: yanli@dhu.edu.cn (Y.L.); wanglu@dhu.edu.cn (L.W.); Tel./Fax: +86-21-67792637 (L.W.)

Received: 20 December 2018; Accepted: 22 February 2019; Published: 1 March 2019

Abstract: Mesh infection is a major complication of hernia surgery after polypropylene (PP) mesh implantation. Modifying the PP mesh with antibacterial drugs is an effective way to reduce the chance of infection, but the hydrophobic characteristic of PP fibers has obstructed the drug adhesion. Therefore, to prepare antimicrobial PP mesh with a stable drug coating layer and to slow the drug release property during the hernia repair process has a great practical meaning. In this work, PP meshes were coated by bio-inspired polydopamine (PDA), which can load and release levofloxacin. PP meshes were activated with cold oxygen plasma and then plasma activated PP fibers were coated with PDA. The PDA coated meshes were further soaked in levofloxacin. The levofloxacin loaded PP meshes demonstrate excellent antimicrobial properties for 6 days and the drug release has lasted for at least 24 h. Moreover, a control PP mesh sample without plasma treatment was also prepared, after coating with PDA and loading levofloxacin. The antimicrobial property was sustained only for two days. The maximum inhibition zone of PDA coated meshes with and without plasma treatment was 12.5 and 9 mm, respectively. On all accounts, the modification strategy can facilely lead to long-term property of infection prevention.

Keywords: polypropylene; hernia meshes; antibacterial; drug release; polydopamine

1. Introduction

An abdominal wall hernia operation is the repositioning of the hernia contents within the abdominal cavity using sutures or mesh materials. Abdominal closure or reinforcement of hernia defects has been performed with numerous mesh materials. Indeed, mesh implantation for hernia repair has reduced hernia recurrence rate, compared to suturing techniques [1]. Synthetic mesh materials have been successful for hernia repair since the last decade. Nevertheless, among all these materials, polypropylene (PP) mesh has been considered as one of the best materials for repairing hernias, owing to its inherent characteristics including that it is inert, hydrophobic, has a strong mechanical property, and is lightweight [2–4]. However, a significant complication of hernia repair is PP mesh induced infection [5,6]. In addition, PP is a hydrophobic polymer which does not absorb drugs during pre-operative prophylaxis [7,8], thereby making the drug coating, to prevent infection, infeasible for PP mesh. Mesh infection can be a serious problem if not solved in the initial stages of mesh implantation. After infection, it is difficult to cure the infected surgical area because heavy antibiotic doses to the body of the patient may cause side effects. Furthermore, the recurrence of the hernia may cause the removal of the hernia mesh, which can be costly in terms of the wealth and health of patients [9,10]. Therefore, creating an active surface of PP mesh and providing the binding sites for drug loading during mesh implantation are the key factors to reduce PP hernia mesh infection [11]. Different kinds of surface modifications of PP meshes have been performed to incorporate different antibiotics for slow drug release [12,13]. Amongst those modification protocols, plasma treatment is considered a crucial step to modify the hydrophobic surfaces of PP fibers [14,15], to incorporate the functional groups, such as hydroxyl or carboxyl, on the surfaces of PP fibers [16], and those functional groups were further utilized to graft host–guest molecules to capture antibiotics [17]. Surface functionalization of PP fibers with cold oxygen plasma is a well-known technique [18,19]. Cold oxygen plasma at low pressure has been especially considered as the most suitable plasma treatment, without changing their bulk properties, for PP fibers [20].

The new advancement in the preparation of medical devices is focused on producing biocompatible materials [21,22]. Thus, dopamine, a bio-inspired material, receives the attention of researchers due to the properties of self-polymerization to polydopamine (PDA) and co-deposition onto surfaces of the materials [23,24]. The PDA coating, or functionalization of material, is a straightforward approach by dipping materials in pH 8.5. The dopamine self-polymerization can be controlled by reaction time or concentration [25,26].

In this work, we prepared antibiotic PP mesh materials functionalized with dopamine via cold O_2 plasma treatment. PP mesh fiber surfaces were activated with oxygen plasma to create hydroxyl or carboxyl groups and then plasma treated meshes were further modified with self-polymerized dopamine for 12–24 h. The PDA incorporated PP meshes were soaked in levofloxacin (0.5%) for 24 h to absorb water-soluble drugs for antimicrobial properties, as shown in Figure 1.

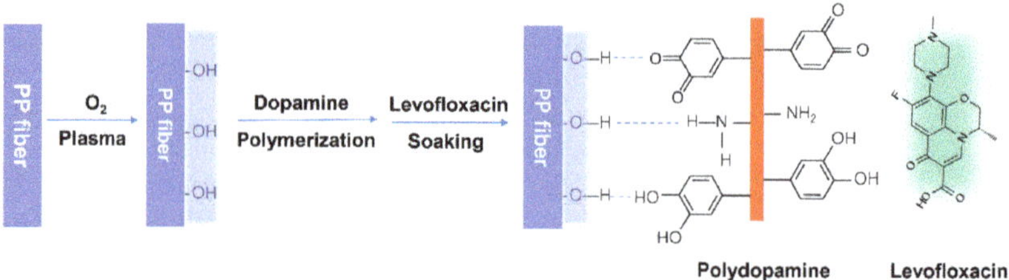

Figure 1. The schematic illustration of PDA coating and subsequent loading levofloxacin on PP meshes.

The PDA coated PP meshes were evaluated for surface morphology, structural changes, antimicrobial properties, and drug release. The results confirmed that dopamine was incorporated on PP fiber surfaces and formed strong interfacial bonding between PP meshes and PDA. Moreover, PDA coated and levofloxacin soaked meshes showed excellent antimicrobial inhibition zones and demonstrated a controlled drug release.

2. Materials and Methods

2.1. Materials

Light weight (27 g/m^2) and large hexagonal pore (sized 3.5 mm × 2.5 mm) monofilament polypropylene mesh devices were received from Nantong Newtec Textile and Chemical Fiber Co., Ltd., Nantong, China. Levofloxacin hydrochloride (98%) was purchased from Energy Chemicals, Shanghai, China. Dopamine and tris(hydroxymethyl)aminomethane (Tris) were received from Aladdin Chemicals Ltd., Shanghai, China.

2.2. Surface Functionalization of PP with Oxygen Plasma

PP mesh samples of equal size (10 cm × 10 cm) were surface activated using an HD-300 cold plasma machine (Changzhou Zhongke Changtai Plasma Technology Co., Ltd., Changzhou, China). The samples were surface functionalized with oxygen plasma using optimized parameters of time (300 s), power (45 W), and pressure (10 Pa). However, flow of oxygen gas was less than 0.2 bars. PP samples after plasma treatment were named as PL-PP.

2.3. Coating of Polydopamine (PDA) on PP Meshes

Dopamine solution was prepared by dissolving 0.1 g of dopamine (2 mg/mL) in 50 mL of Tris-HCl buffer (10 mM, pH 8.5) [25]. The liquor ratio of the PP meshes to the solution was 1:100. Dopamine self-polymerization (non-plasma and oxygen plasma treated samples) was performed at a temperature 37 °C, with continuous stirring (80 rpm) for 12–24 h. After 12–24 h samples were taken out, rinsed several times with hot (50 °C) and cold deionized water, and dried at 50 °C in a vacuum oven. All samples were coated with the same concentration, the only difference was time. PDA coated samples after plasma treatment for 12 and 24 h were named PL-Dop-12 and PL-Dop-24, respectively. Non-plasma treated samples of coated PDA were named Dop-12 and Dop-24.

2.4. Loading and Release of Levofloxacin

The PDA coated PP (0.5 g) meshes were soaked for 24 h in a 50 mL solution containing 0.3 g (0.6%) levofloxacin. All samples were soaked in same concentration of levofloxacin for the same soaking time of 24 h. After 24 h, loading samples were taken out and dried at 40 °C. Thus, after levofloxacin, loaded samples were named Dop-LVFX-24, PL-Dop-LVFX-12, and PL-Dop-LVFX-24.

The drug release profiles of levofloxacin loaded samples (2 cm × 2 cm) were measured in a centrifugal tube containing 8 mL of phosphate-buffered saline (PBS) and the PBS mixture with mesh samples was shaken the for required duration at 70 rpm. The UV-spectrophotometry (TU-1901 Beijing Purkinjie Co., Ltd., Beijing, China) was used at 37 °C for absorption measurements. During each absorption measurement, 1 mL of the mixture solution of each sample was extracted and 1 mL of fresh PBS was added into the mixture. Absorption of all samples was measured at 290 nm. The accumulative levofloxacin release (%) was obtained as the ratio of levofloxacin release to the total drug loaded onto the samples. Average absorption values were used to calculate the levofloxacin release (%).

2.5. Characterization of Modified PP Meshes

PDA coated and levofloxacin loaded samples were coated with platinum (Pt) and observed with a Scanning Electron Microscope (SEM) (Quanta SEM 250, FEI™, Waltham, MA, USA). Moreover, an Energy Dispersive X-ray spectrometer (EDX) (ISIS 300, Oxford Instruments, Oxfordshire, UK) was connected with the SEM for element analysis of the PDA-modified and untreated PP samples. Fourier Transform Infrared Spectroscopy (FTIR) (Nicolet 6700, Thermo Fisher Scientific, Waltham, MA, USA) of Attenuated Total Reflection (ATR) mode was used to characterize all modified and control samples. All samples were analyzed in the wavenumber range of 500–4000 cm^{-1} with resolution of 4.0 cm^{-1}. An atomic force microscope (AFM) (Technologies 5500, Keysight, Santa Rosa, CA, USA) was used to analyze the roughness of the different surfaces.

2.6. Antibacterial Properties

Qualitative Analysis

The agar diffusion test method was performed to assess the antibacterial properties of the dopamine coated and levofloxacin loaded samples described in our recently published paper in *Polymers* [27]. 400 µL of 1×10^8 CFU/mL bacterial suspension was spread on agar plates. Sterilized swabs were used to evenly spread the bacteria. Two types of bacteria namely *Escherichia* (*E.C*) (ATCC®

25922™, Shanghai, China) and *Staphylococcus aureus* (*S.A*) (ATCC® 25923™, Shanghai, China) were used for antibacterial analysis of the modified meshes. Modified and untreated samples (1 cm × 1 cm square) were placed in the center of agar plates containing bacteria (*S.A* and *E.C*). The agars plated with samples were incubated in a standard oven, maintaining 37 °C for 24 h. The zone of inhibition of each sample was measured with a digital vernier caliper and all results from the inhibition zones were measured in 4 directions and reported as the average values. The zone of inhibition was measured using the following formula: $X = (Y - Z)/2$, where X = the inhibition zone, Y = the inhibition zone measured after 24 h incubation, and Z = the inhibition zone before incubation.

Furthermore, we observed the antibacterial properties of the modified and untreated PP samples in a controlled environment of 37 °C. After each day, samples were taken out of the oven and transferred to fresh agar plates containing bacteria. The zone of inhibition was measured to see the change in antibacterial properties according to the number of days.

3. Results and Discussion

3.1. Polydopamine Coated and Levofloxacin Loaded

Polypropylene (PP) warp knitted meshes were soaked in a weak alkaline solution of dopamine. It was expected that dopamine (PDA) would self-polymerize and co-deposit onto PP meshes. Two steps were performed in the surface modification of PP meshes. In the first step, PP meshes were activated with oxygen plasma and in the second step more PDA was adhered to oxygen plasma activated PP meshes via hydrogen bond.

Plasma treated and untreated PP meshes were coated in the same concentration of dopamine and at the same temperature (37 °C) for 12 and 24 h and we received a marginal difference between oxygen plasma treated and untreated coating amounts. The weight increase of oxygen plasma treated and untreated meshes, for 12 and 24 h, is shown in Figure 2. The corresponding weight of the Dop-12 and Dop-24 is 0.3% and 0.5%, respectively. However, the average weight increase of PL-Dop-12 and PL-Dop-24 is 2.81% and 3.75%, respectively. Moreover, after 24 to 36 h soaking times there was no change in the amount of dopamine coating.

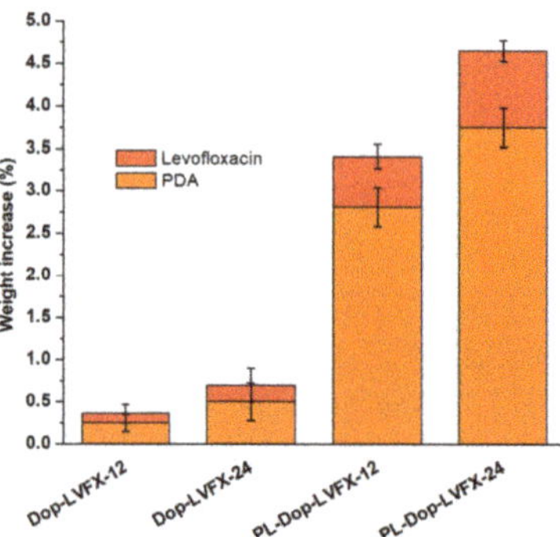

Figure 2. The weight increase (%) of untreated and plasma treated PP meshes after PDA and levofloxacin coating.

All modified PP meshes were soaked in the same concentration (0.6%) of levofloxacin for 24 h. Plasma treated and PDA coated PP meshes were weighed before and after loading levofloxacin. From Figure 2, PL-Dop-LVFX-12 and PL-Dop-LVFX-24 increase the corresponding weights by 0.6% and 0.9%, respectively. Moreover, samples without plasma treatment (Dop-LVFX-12 and Dop-LVFX-24) increase corresponding weight by only 0.2%. Therefore, non-plasma treated samples (Dop-24, Dop-LVFX-24) and plasma treated samples (PL-Dop-LVFX-12 and PL-Dop-LVFX-24) were prepared for further characterization.

3.2. Surface Morphology of Modified PP Meshes

Surface roughness of PP control and PL-PP meshes were observed using an AFM (Figure 3). Prior to oxygen plasma activation the surface of untreated mesh fibers is smooth with obvious mark lines. After oxygen plasma modification, the surface of PP fibers becomes rougher and the surface piles are more visible, instead of line marks. Moreover, after the oxygen plasma treatment, S_q and S_a parameters of the modified meshes are increased by 55.5% and 21.2%, respectively, showing a rougher surface as compared to the PP control fibers. Therefore, oxygen plasma functionalization significantly changes the surface of PP meshes, which is consistent with the literature [28].

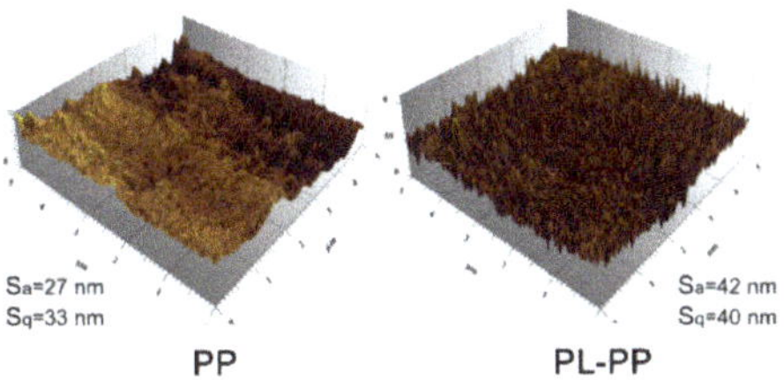

Figure 3. AFM of PP control and PL-PP.

Figure 4 shows SEM images of the modified PP samples. It can be observed that the PP control displays identical sharp and thin lines before oxygen plasma treatment (Figure 4a1,a2). Nevertheless, after plasma treatment a fuzzy and slightly dark appearance, without cut and lines (Figure 4b1,b2) shows a rougher surface of PP mesh. Moreover, for Dop-24 (Figure 4c1,c2), PDA can be observed on the surfaces of fibers with small beads, but PP mesh fibers are still not fully covered with PDA. On the other hand, PL-Dop-12 (Figure 4d1,d2) shows a more PDA coated area of PP fibers. This may be because, after plasma treatment, −OH groups on the PP surface enhanced the adherence of PDA more effectively via hydrogen bond, compared to non-plasma treated meshes. Moreover, PL-Dop-24 (Figure 4e1,e2) shows remarkably increased coating with wider beads, which covered whole spherical fiber area.

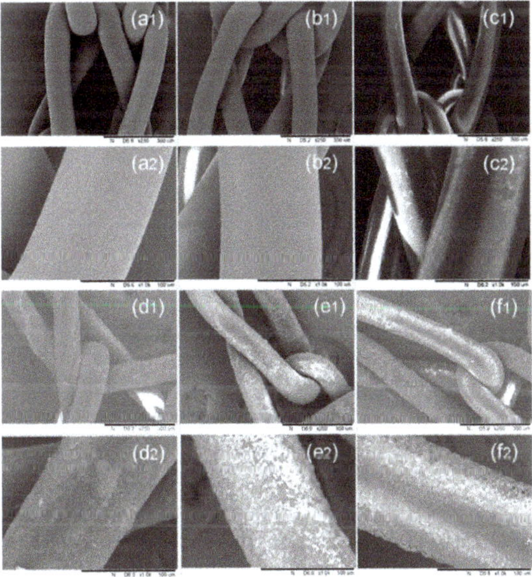

Figure 4. SEM images as follows: PP control (**a1,a2**), PL-PP (**b1,b2**), Dop-24 (**c1,c2**), PL-Dop-12 (**d1,d2**), PL-Dop-24 (**e1,e2**), and PL-Dop-LVFX-24 (**f1,f2**). Top row (**a1–f1**) ×250 and bottom row (**a2–f2**) ×1000.

Furthermore, PL-Dop-LVFX-24 (Figure 4f1,f2) shows a more swollen and brighter structure, which may be due to levofloxacin absorbed by PDA coated meshes via catechol and hydrogen bond. Above all, plasma treated meshes could be effectively coated with PDA. This is due to the plasma surface modification of PP fibers, which increases surface wettability [29] as well as surface adhesion [30] for PDA coating.

3.3. Chemical Characterization of Modified PP Meshes

Figure 5 displays the element analysis of PP control and modified meshes using Energy Despersive EDX. Figure 5a shows only the carbon (C) atom peak within 0.5 keV, which confirms PP fibers. PL-PP meshes show additional oxygen atom (O), which confirms oxygen-containing groups, generated on the surface of PP meshes after oxygen plasma treatment, as shown in Figure 5b. Furthermore, Figure 5c displays carbon (C), oxygen (O), and an additional nitrogen (N) atom between the oxygen and carbon peaks, which confirms that the PP meshes were coated by PDA and that the oxygen atoms' weight (%) increased from 2.85% to 7.89%. Moreover, after levofloxacin loading, the presence of fluorine (F) atom is displayed within 0.6 keV (Figure 5d). Thus, it proves the existence of levofloxacin on the surface of PDA coated PP meshes. These results are similar to our recent published paper, except for a different weight (%) of each element [27].

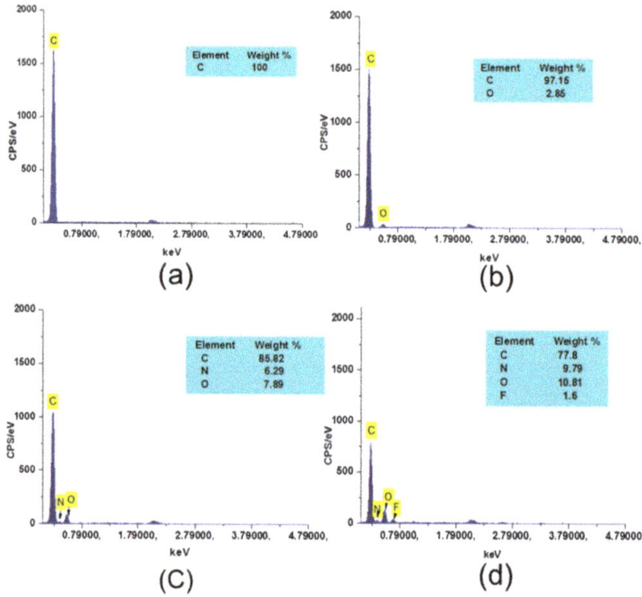

Figure 5. EDX spectra and weight of element (%) (**a**) PP control; (**b**) PL-PP; (**c**) PL-Dop-24; and (**d**) PL-Dop-LVFX-24.

FTIR spectra of untreated and treated samples are shown in Figure 6. The PP control shows peaks at 2950, 2916, 1452, and 1376 cm^{-1} [17]. Thus, except for the PP control, PL-PP meshes display additional peaks at 3264 cm^{-1}, which is due to the hydroxyl ($-$OH) on the surface of PP meshes. Moreover, it can be observed that Dop-24 shows slight vibration peaks at 1626 and 1532 cm^{-1}, which is due to the establishment of ν C=O and δ NH, respectively. Furthermore, C–H (ν CH) stretch can be observed at 1290 cm^{-1}. PL-Dop-24 shows similar peaks of 1626 and 1532 cm^{-1}. More intensity of the absorbance band is found, which shows hydrogen bonding between PDA and PP meshes. Moreover, PL-Dop-LVFX-24 shows similar peaks as PDA coating peaks because levofloxacin was just loaded on the layer and did not react with PDA.

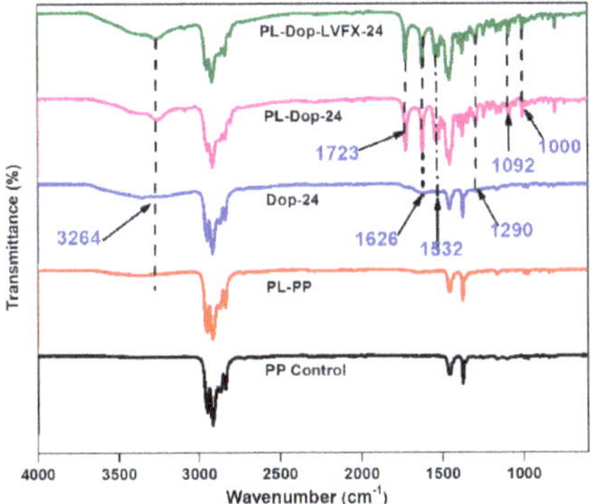

Figure 6. FTIR spectra (ATR) of PP control, PL-PP, Dop-24, PL-Dop-24, and PL-Dop-LVFX-24.

3.4. Antibacterial Activity

Antibacterial activity of PP meshes at different stages of modification was performed by the agar diffusion method, as shown in Figure 7. It can be observed that the PP control did not display antibacterial properties. Nevertheless, PP meshes activated with oxygen plasma have slight antibacterial properties. Dop-LVFX-24 showed good antibacterial properties. The average inhibition zone for *E.C* and *S.A* was 9.6 and 8.5 mm, respectively. With plasma treatment, PL-Dop-LVFX-12 performed better and showed an average inhibition zone of 12.5 and 12 mm for *E.C* and *S.A*, respectively. PL-Dop-LVFX-24 demonstrated excellent antibacterial properties with an inhibition zone of 15 and 14.5 mm for *E.C* and *S.A*, respectively.

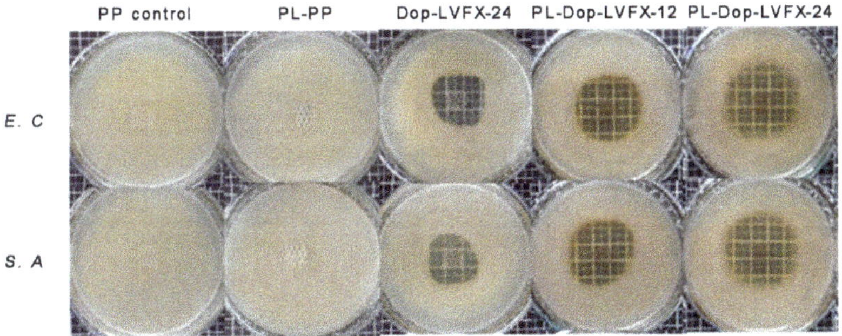

Figure 7. Inhibition zone diameter of PP control, PL-PP, Dop-LVFX-24, PL-Dop-LVFX-12, and PL-Dop-LVFX-24. All samples in the top row = *E.C* and the bottom row = *S.A*.

Moreover, antibacterial release properties of modified PP meshes (Figure 8) were determined after each 24 h. It can be observed that Dop-LVFX-24 exhibited antibacterial properties for only two days, as shown in Figure 8a. This may because less levofloxacin was adhered on the surface due to less PDA coating. However, PDA coated PP meshes, after plasma treatment for 12 h and soaked in the levofloxacin (Dop-LVFX-12), showed good antibacterial properties for 5 days, with a reduced regular inhibition zone throughout 5 days. The minimum inhibition zone on 5th day was 3 and 2.5 mm for *E.C* and *S.A*, respectively (Figure 8b). A bigger average inhibition zone was displayed by samples using *E.C* during the entire 5 days. Moreover, PDA coated meshes, for 24 h and soaked in levofloxacin (Dop-LVFX-24), demonstrated a more sustained antibacterial release for 6 days (Figure 8c). It was also observed that the inhibition zone of *E.C* was greater, in comparison to *S.A*, for an entire 6 days, but on 4th day almost the same inhibition zone was shown for the two types of bacteria. The minimum inhibition zone for *E.C* and *S.A* was 2.99 and 2.05 mm respectively.

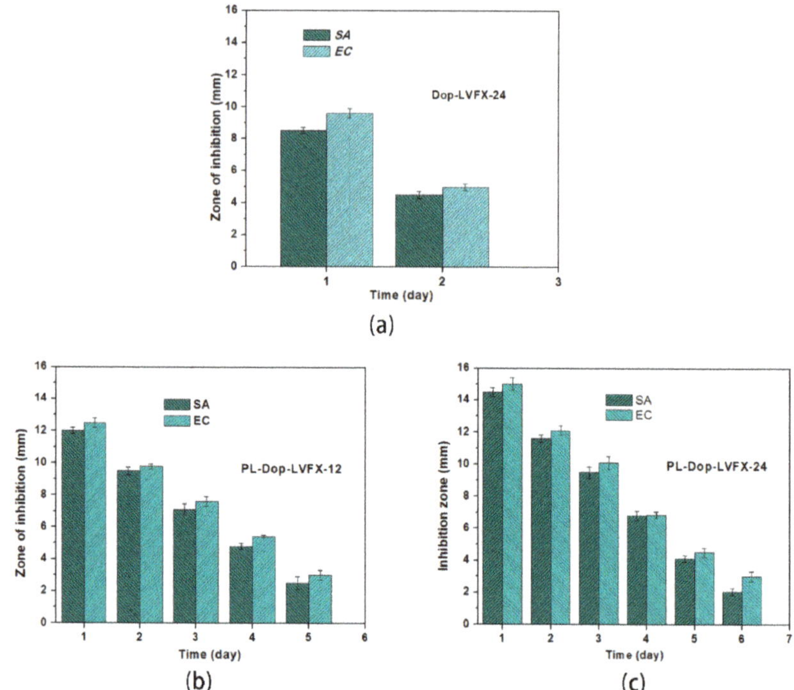

Figure 8. Antibacterial release properties of (**a**) Dop-LVFX-24, (**b**) PL-Dop-LVFX-12, and (**c**) PL-Dop-LVFX-24.

In our previous work, we prepared PP meshes with a one-step coating of PDA, without plasma treatment [31]. Herein, we found better results with PDA coating using oxygen plasma. Thus, we compared antibacterial performance of PDA coating without plasma and PDA coating with oxygen plasma. Overall, plasma treated and PDA coated PP meshes, PL-Dop-LVFX-12 and PL-Dop-LVFX-24, displayed more sustained antibacterial properties than non-plasma treated PP meshes (Dop-LVFX-24). This may because plasma treatment improved the coating efficiency of PDA onto PP fibers with increased corresponding weight and strong bonding, which may hold more levofloxacin and release antibacterial properties for a longer duration of time compared to non-plasma treated PP meshes.

3.5. Drug Release Profile of PDA Coated and Levofloxacin Loaded PP Meshes

The average accumulative drug release (%) of three samples (Dop-LVFX-24, PL-Dop-LVFX-12, and PL-Dop-LVFX-24) was performed in a neutral environment using a PBS solution. From Figure 9 and Table 1, it can be observed that Dop-LVFX-24 showed fast drug release and about 100% of the drug was released within first 6 h, though a major portion (80%) of the drug was released within 3 h. Moreover, PL-Dop-LVFX-12 displayed a sustained drug release profile. In first 3 h, 65% of drug was released; after 6 h, the accumulative drug release was 78.8%; and after 12 h, 91.9% was released. Further sustained drug release (99.1%) continued up to 24 h. Moreover, PL-Dop-LVFX-24 displayed more sustained levofloxacin release compared to PL-Dop-LVFX-12. In the first 3 h, an accumulative 60.88% of the drug was released. After 6 h, 72.2% of levofloxacin was released. At 12 h, 86.3% of the drug was released and a total of 93.3% of drug was released at 24 h. Thus, PL-Dop-LVFX-24 samples showed sustained accumulative drug release. Drug releases of all three samples can explain antibacterial activity performance. The PL-Dop-LVFX-24 sample could be a suitable product and may be used for hernia mesh implantation.

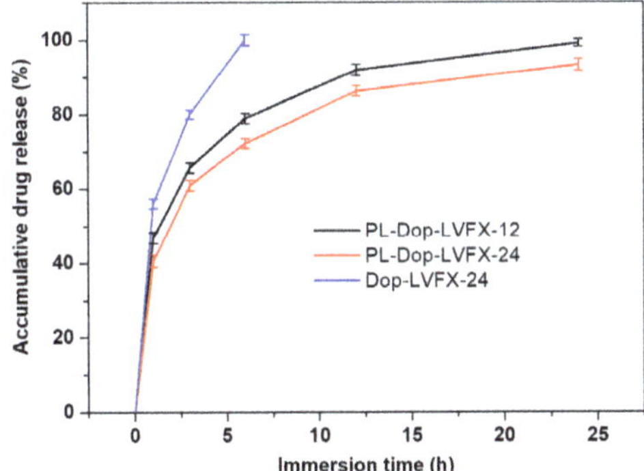

Figure 9. Drug release curves of PDA coated and levofloxacin loaded PP meshes.

Table 1. The average accumulative drug release of three samples.

Time (h)	Dop-LVFX-24 (%)	PL-Dop-LVFX-12 (%)	PL-Dop-LVFX-24 (%)
0	0	0	0
1	56 ± 1.3	47 ± 1.5	41 ± 1.6
3	80 ± 1.2	65 ± 1.3	61 ± 1.5
6	100 ± 1.5	79 ± 1.3	72 ± 1.3
12	–	92 ± 1.5	86 ± 1.4
24	–	99 ± 1.0	93 ± 1.6

In our previous study, cyclodextrins were used to capture levofloxacin HCL for antibacterial and drug release properties. The release of the drug was observed for at least 24 h [32]. Herein, polydopamine (PDA) coated PP meshes were soaked in the levofloxacin HCl solution and modified PP meshes showed a better antibacterial inhibition zone than PP meshes modified with cyclodextrins, but cyclodextrins had released the drug for a longer duration of time than polydopamine. The reason for this could be that cyclodextrins have cavities which releases the drug slowly.

Moreover, commonly used drug carriers, such as cyclodextrins [33–35] and chitosan [36,37], may release drugs for longer durations of time, compared to polydopamine, but the inhibition zones may be smaller. Overall, PDA modified PP meshes released levofloxacin HCL for 24 h and showed excellent antibacterial inhibition zones of 15 and 14.5 mm for *E.C* and *S.A*, respectively. Considering the burst release of drugs, a hernia mesh infection is necessary to control in the initial stages of mesh implantation [32,38]. Thus, the PDA and levofloxacin HCL coated PP meshes with good drug release properties could be useful for prevention of hernia mesh infections.

4. Conclusions

Bio-inspired PDA coated antimicrobial PP meshes for hernia repair were successfully prepared. The oxygen plasma treated PP meshes show better results of PDA coating than the raw one. It is proved that PDA coated PP meshes, after oxygen plasma treatment, can absorb more levofloxacin and release the drug for a longer suitable time, which demonstrates reasonable antimicrobial release properties. Thus, plasma treated PP meshes coated with PDA and loaded with levofloxacin may be a good choice for prevention of hernia mesh infections.

Author Contributions: Conceptualization X.S., N.S. and Y.L.; Methodology X.S. and Y.Q.; Software, N.S.; Validation, A.K. and X.S.; Writing Original Draft, L.W. and J.G.; Writing–Review and Editing, G.B. and R.G.

Funding: The Fundamental Research Funds for the Central Universities (No. 17D110111 & 2232018G-01), the National Key Research Development Program of China (No. 2016YFB 0303300), and 111 project "Biomedical Textile Material Science and Technology" (No. B07024).

Acknowledgments: Thanks to students of the Biomedical Textile Materials Group for their help during experiments and testing.

Conflicts of Interest: The authors declare no conflict of interest.

References

1. Guillaume, O.; Perez-Tanoira, R.; Fortelny, R.; Redl, H.; Moriarty, T.F.; Richards, R.G.; Eglin, D.; Petter Puchner, A. Infections associated with mesh repairs of abdominal wall hernias: Are antimicrobial biomaterials the longed-for solution? *Biomaterials* **2018**, *167*, 15–31. [CrossRef] [PubMed]
2. Greca, F.; Paula, J.; Biondo-Simões, M.; Costa, F.; Silva, A.; Time, S.; Mansur, A. The influence of differing pore sizes on the biocompatibility of two polypropylene meshes in the repair of abdominal defects. *Hernia* **2001**, *5*, 59–64. [PubMed]
3. Hazebroek, E.J.; Ng, A.; Yong, D.H.; Berry, H.; Leibman, S.; Smith, G.S. Evaluation of lightweight titanium-coated polypropylene mesh (TiMesh) for laparoscopic repair of large hiatal hernias. *Surg. Endosc.* **2008**, *22*, 2428–2432. [CrossRef] [PubMed]
4. Jerabek, J.; Novotny, T.; Vesely, K.; Cagas, J.; Jedlicka, V.; Vlcek, P.; Capov, I. Evaluation of three purely polypropylene meshes of different pore sizes in an onlay position in a New Zealand white rabbit model. *Hernia* **2014**, *18*, 855–864. [CrossRef] [PubMed]
5. Kulaga, E.; Ploux, L.; Balan, L.; Schrodj, G.; Roucoules, V. Mechanically responsive antibacterial plasma polymer coatings for textile biomaterials. *Plasma Process. Polym.* **2014**, *11*, 63–79. [CrossRef]
6. Knetsch, M.L.W.; Koole, L.H. New strategies in the development of antimicrobial coatings: The example of increasing usage of silver and silver nanoparticles. *Polymers* **2011**, *3*, 340–366. [CrossRef]
7. Mazaki, T.; Mado, K.; Masuda, H.; Shiono, M.; Tochikura, N.; Kaburagi, M. A randomized trial of antibiotic prophylaxis for the prevention of surgical site infection after open mesh-plug hernia repair. *Am. J. Surg.* **2014**, *207*, 476–484. [CrossRef] [PubMed]
8. Mazaki, T.; Mado, K.; Masuda, H.; Shiono, M. Antibiotic prophylaxis for the prevention of surgical site infection after tension-free hernia repair: A Bayesian and frequentist meta-analysis. *J. Am. Coll. Surg.* **2013**, *217*, 788–801. [CrossRef] [PubMed]
9. Perez-Kohler, B.; Fernandez-Gutierrez, M.; Pascual, G.; Garcia-Moreno, F.; San Roman, J.; Bellon, J.M. In vitro assessment of an antibacterial quaternary ammonium-based polymer loaded with chlorhexidine for the coating of polypropylene prosthetic meshes. *Hernia* **2016**, *20*, 869–878. [CrossRef] [PubMed]
10. Zhang, Z.; Tang, J.; Wang, H.; Xia, Q.; Xu, S.; Han, C.C. Controlled antibiotics release system through simple blended electrospun fibers for sustained antibacterial effects. *ACS Appl. Mater. Interfaces* **2015**, *7*, 26400–26404. [CrossRef] [PubMed]
11. Sanbhal, N.; Miao, L.; Xu, R.; Khatri, A.; Wang, L. Physical structure and mechanical properties of knitted hernia mesh materials: A review. *J. Ind. Text.* **2017**, *48*, 333–360. [CrossRef]
12. Muzio, G.; Miola, M.; Perero, S.; Oraldi, M.; Maggiora, M.; Ferraris, S.; Vernè, E.; Festa, V.; Festa, F.; Canuto, R.A.; et al. Polypropylene prostheses coated with silver nanoclusters/silica coating obtained by sputtering: Biocompatibility and antibacterial properties. *Surf. Coat. Technol.* **2017**, *319*, 326–334. [CrossRef]
13. Labay, C.; Canal, J.M.; Modic, M.; Cvelbar, U.; Quiles, M.; Armengol, M.; Arbos, M.A.; Gil, F.J.; Canal, C. Antibiotic-loaded polypropylene surgical meshes with suitable biological behaviour by plasma functionalization and polymerization. *Biomaterials* **2015**, *71*, 132–144. [CrossRef] [PubMed]
14. Nisticò, R.; Rosellini, A.; Rivolo, P.; Faga, M.G.; Lamberti, R.; Martorana, S.; Castellino, M.; Virga, A.; Mandracci, P.; Malandrino, M.; et al. Surface functionalisation of polypropylene hernia-repair meshes by RF-activated plasma polymerisation of acrylic acid and silver nanoparticles. *Appl. Surf. Sci.* **2015**, *328*, 287–295. [CrossRef]
15. Nisticò, R.; Magnacca, G.; Faga, M.G.; Gautier, G.; D'Angelo, D.; Ciancio, E.; Lamberti, R.; Martorana, S. Effect of atmospheric oxidative plasma treatments on polypropylenic fibers surface: Characterization and reaction mechanisms. *Appl. Surf. Sci.* **2013**, *279*, 285–292. [CrossRef]
16. Zhang, Z.; Zhang, T.; Li, J.; Ji, Z.; Zhou, H.; Zhou, X.; Gu, N. Preparation of poly(L-lactic acid)-modified polypropylene mesh and its antiadhesion in experimental abdominal wall defect repair. *J. Biomed. Mater. Res. Part. B Appl. Biomater.* **2014**, *102*, 12–21. [CrossRef] [PubMed]

17. Sanbhal, N.; Mao, Y.; Sun, G.; Xu, R.F.; Zhang, Q.; Wang, L. Surface modification of polypropylene mesh devices with cyclodextrin via cold plasma for hernia repair: Characterization and antibacterial properties. *Appl. Surf. Sci.* **2018**, *439*, 749–759. [CrossRef]
18. Jelil, R.A. A review of low-temperature plasma treatment of textile materials. *J. Mater. Sci.* **2015**, *50*, 5913–5943. [CrossRef]
19. Lai, J.; Sunderland, B.; Xue, J.; Yan, S.; Zhao, W.; Folkard, M.; Michael, B.D.; Wang, Y. Study on hydrophilicity of polymer surfaces improved by plasma treatment. *Appl. Surf. Sci.* **2006**, *252*, 3375–3379. [CrossRef]
20. Khelifa, F.; Ershov, S.; Habibi, Y.; Snyders, R.; Dubois, P. Free-radical-induced grafting from plasma polymer surfaces. *Chem. Rev.* **2016**, *116*, 3975–4005. [CrossRef] [PubMed]
21. Patel, H.; Ostergard, D.R.; Sternschuss, G. Polypropylene mesh and the host response. *Int. Urogynecol. J.* **2012**, *23*, 669–679. [CrossRef] [PubMed]
22. Perez-Kohler, B.; Bayon, Y.; Bellon, J.M. Mesh infection and hernia repair: A review. *Surg. Infect.* **2016**, *17*, 124–137. [CrossRef] [PubMed]
23. Liu, Y.; Fang, Y.; Qian, J.; Liu, Z.; Yang, B.; Wang, X. Bio-inspired polydopamine functionalization of carbon fiber for improving the interfacial adhesion of polypropylene composites. *RSC Adv.* **2015**, *5*, 107652–107661. [CrossRef]
24. Ngo, T.H.A.; Nguyen, D.T.; Do, K.D.; Nguyen, T.T.M.; Mori, S.; Tran, D.T. Surface modification of polyamide thin film composite membrane by coating of titanium dioxide nanoparticles. *J. Sci. Adv. Mater. Devices* **2016**, *1*, 468–475. [CrossRef]
25. Ying, T.; Guo-Xin, T.; Cheng-Yun, N.; Xi-Cang, R.; Yu, Z.; Lei, Z. Bioinspired polydopamine functionalization of titanium surface for silvernanoparticles immobilization with antibacterial property. *J. Inorg. Mater.* **2014**, *29*, 1320–1326. [CrossRef]
26. Zhang, R.X.; Breaken, L.; Liu, T.Y.; Luis, P.; Wang, X.L.; Van der Bruggen, B. Remarkable anti-fouling performance of TiO$_2$-modified TFC membranes with mussel-inspired polydopamine binding. *Appl. Sci.* **2017**, *7*, 81. [CrossRef]
27. Sanbhal, N.; Mao, Y.; Sun, G.; Li, Y.; Peerzada, M.; Wang, L. Preparation and characterization of antibacterial polypropylene meshes with covalently incorporated β-cyclodextrins and captured antimicrobial agent for Hernia repair. *Polymers* **2018**, *10*, 58. [CrossRef]
28. Vishnuvarthanan, M.; Rajeswari, N. Effect of mechanical, barrier and adhesion properties on oxygen plasma surface modified PP. *Innov. Food Sci. Emerg. Technol.* **2015**, *30*, 119–126. [CrossRef]
29. Fauland, G.; Constantin, F.; Gaffar, H.; Bechtold, T. Production scale plasma modification of polypropylene baselayer for improved water management properties. *J. Appl. Polym. Sci.* **2015**, *132*, 41294. [CrossRef]
30. Sorrentino, L.; Carrino, L.; Napolitano, G. Oxygen cold plasma treatment on polypropylene: Influence of process parameters on surface wettability. *Surf. Eng.* **2013**, *23*, 247–252. [CrossRef]
31. Sanbhal, N.; Saitter, X.; Peerzada, M.; Habboush, A.; Wang, F.; Wang, L. One-step surface functionalized hydrophilic polypropylene meshes for hernia repair using bio-inspired polydopamine. *Fibres* **2019**, *7*, 6. [CrossRef]
32. Sanbhal, N.; Saitaer, X.; Li, Y.; Mao, Y.; Zou, T.; Sun, G.; Wang, L. Controlled levofloxacin release and antibacterial properties of β-cyclodextrins-grafted polypropylene mesh devices for hernia repair. *Polymers* **2018**, *10*, 493. [CrossRef]
33. Laurent, T.; Kacem, I.; Blanchemain, N.; Cazaux, F.; Neut, C.; Hildebrand, H.F.; Martel, B. Cyclodextrin and maltodextrin finishing of a polypropylene abdominal wall implant for the prolonged delivery of ciprofloxacin. *Acta Biomater.* **2011**, *7*, 3141–3149. [CrossRef] [PubMed]
34. Otero-Espinar, F.J.; Torres-Labandeira, J.J.; Alvarez-Lorenzo, C.; Blanco-Méndez, J. Cyclodextrins in drug delivery systems. *J. Drug Deliv. Sci. Technol.* **2010**, *20*, 289–301. [CrossRef]
35. Thatiparti, T.R.; Shoffstall, A.J.; von Recum, H.A. Cyclodextrin-based device coatings for affinity-based release of antibiotics. *Biomaterials* **2010**, *31*, 2335–2347. [CrossRef] [PubMed]
36. Avetta, P.; Nisticò, R.; Faga, M.G.; D'Angelo, D.; Boot, E.A.; Lamberti, R.; Martorana, S.; Calza, P.; Fabbri, D.; Magnacca, G. Hernia-repair prosthetic devices functionalised with chitosan and ciprofloxacin coating: Controlled release and antibacterial activity. *J. Mater. Chem. B* **2014**, *2*, 5287–5294. [CrossRef]

37. Nisticò, R.; Faga, M.G.; Gautier, G.; Magnacca, G.; D'Angelo, D.; Ciancio, E.; Piacenza, G.; Lamberti, R.; Martorana, S. Physico-chemical characterization of functionalized polypropylenic fibers for prosthetic applications. *Appl. Surf. Sci.* **2012**, *258*, 7889–7896. [CrossRef]
38. Erdas, E.; Medas, F.; Pisano, G.; Nicolosi, A.; Calo, P.G. Antibiotic prophylaxis for open mesh repair of groin hernia: Systematic review and meta-analysis. *Hernia* **2016**, *20*, 765–776. [CrossRef] [PubMed]

Article

Designing a Laboratory Bioassay for Evaluating the Efficacy of Antifouling Paints on *Amphibalanus amphitrite* Using a Flow-Through System

Ryuji Kojima [1,*], Seiji Kobayashi [2], Kiyotaka Matsumura [3], Cyril Glenn Perez Satuito [4], Yasuyuki Seki [5], Hirotomo Ando [1] and Ichiro Katsuyama [2]

[1]   Department of Marine Environment and Engine System, National Maritime Research Institute, Mitaka, Tokyo 181-0004, Japan; ando@nmri.go.jp

[2]   Department of Environmental Risk Consulting, Japan NUS Co., Ltd., Shinjuku, Tokyo 166-0023, Japan; kobayasi@janus.co.jp (S.K.); katuyama@janus.co.jp (I.K.)

[3]   School of Marine Bioscience, Kitasato University, Sagamihara, Kanagawa 252-0373, Japan; matsumurasipc@gmail.com

[4]   Graduate School of Fisheries and Environmental Sciences, Nagasaki University, Nagasaki 852-8131, Japan; satuito@nagasaki-u.ac.jp

[5]   Hiroshima R&D Centre, Chugoku Marine Paints, Ltd., Otake, Hiroshima 739-0652, Japan; yasuyuki_seki@cmp.co.jp

*   Correspondence: kojima@nmri.go.jp; Tel.: +81-422-41-3769

Received: 3 December 2018; Accepted: 1 February 2019; Published: 12 February 2019

Abstract: With the aim of establishing a protocol for evaluating the efficacy of antifouling paints on different organisms, a flow-through laboratory test using triangular boxes was developed for cyprids of the barnacle *Amphibalanus* (=*Balanus*) *amphitrite*. Six different formulations of antifouling paints were prepared in increasing content (0 to 40 wt.%) of Cu_2O, which is the most commonly used antifouling substance, and each formulation of paint was coated on one surface of each test plate. The test plates were aged for 45 days by rotating them at a speed of 10 knots inside a cylinder drum with continuously flowing seawater. The settlement behavior of 3-day-old cyprids released inside triangular boxes made from the test plates was observed. A decreasing number of juveniles settled on surfaces of test plates that were coated with paint containing more than 30 wt.% of Cu_2O. Results of the laboratory bioassays were consistent with those from the field experiments.

Keywords: antifouling efficacy; flow-through; triangular box; *Amphibalanus amphitrite*; cuprous oxide; dynamic aging; repellant activity; raft experiment; bioassay; biofouling of ships' hull

---

1. Introduction

A wide range of macrofoulers have been used as test organisms in antifouling bioassays conducted under controlled experimental conditions [1,2]. Barnacles are typical fouling organisms that attach to ships' hulls and submerged artificial structures. This biofouling consequently leads to increased fuel consumption and accidental introduction of non-indigenous species to another marine environment, possibly causing significant and harmful changes [3–12]. Controlling the attachment of barnacles is of great significance in the development of antifouling technology.

Internally fertilized eggs of barnacles' hatch when embryos develop in the mantle cavity of brooding adults, and they then start to drift in the ocean. During the planktonic larval stage, barnacles molt through six naupliar stages before they molt to cyprids that do not feed. Cyprids attach to a suitable substrate and metamorphose into juveniles to start a sessile life stage. This process of attachment and metamorphosis is most important in the life cycle of barnacles. The inhibition of this process is the key to the development of antifouling technology on barnacles.

Many laboratory studies using cyprids have been conducted to elucidate the settlement mechanism of barnacles and to search for anti-foulants [13–19]. The barnacle *Amphibalanus amphitrite* (*A. amphitrite*) is widely distributed in the intertidal zones of the subtropical and temperate regions of the globe [16–18]. The establishment of its larval rearing method has made its cyprids available in the laboratory all year round [16]. They are also used worldwide as a model species in larval settlement studies of barnacles [20,21]. Most studies have been carried out under a static condition [20,22–31]. Nevertheless, antifouling agents are designed to be slowly released. This means that in an evaluation test for antifouling efficacy, the accumulation of the antifouling agent becomes a huge concern under a static condition. Therefore, the ideal assay condition should be a flow-through water system that does not allow the accumulation of antifouling agents in the test water. Re-circulating the test water inside the tank was proposed as a new settlement assay method to address problems encountered under a static water condition [18,32].

A new flow-through bioassay was reported by Pansch et al. [15] as a tool for rapid laboratory-based screening of candidate compounds for use in antifouling coatings. In a previous investigation, the authors designed a bioassay with a flow-through water system and successfully assessed the efficacy of antifouling paints using the mussel—*Mytilus galloprovincialis* (*M. galloprovincialis*) [33]. The flow-through bioassay system designed for *M. galloprovincialis* was unique in that it was compact and cost-effective, since bioassay vessels that were used were small and required a lesser amount of seawater during the test. In order to comprehensively assess the efficacy of antifouling paints, it is important to conduct laboratory bioassays on more than one test organism, since sensitivity to biocide may vary among fouling species. Therefore, the validation of laboratory bioassays for assessing the efficacy of antifouling paints will require broadening the objectivity of the laboratory experiment by including other fouling organisms. Hence, the authors designed a flow-through bioassay system for *A. amphitrite* that incorporated the features of our previous system.

In this study, the authors proposed and validated a newly developed laboratory bioassay for evaluating the efficacy of biocide-releasing antifouling paints with in a flow-through system using *A. amphitrite*. In order to conduct this study, test paints containing the antifoulant cuprous oxide (Cu_2O) were prepared in varying concentrations ranging from 0 to 40 wt.%. To simulate the actual condition of ship hulls, test plates coated with the test paints were initially cured dynamically and the inhibition effects of the paints was evaluated using cyprid settlement as the index. A comparison of results from laboratory bioassays using barnacle larvae and from field experiments conducted on rafts is also discussed.

2. Materials and Methods

2.1. AF Paints and Test Plates

Six types of AF paints with vinyl copolymer coatings and containing 0, 5, 10, 20, 30, and 40 wt.% of Cu_2O were prepared, as shown in Table 1. Polyvinyl chloride (PVC) plates used in laboratory experiments were 50 mm $\times$ 50 mm $\times$ 2 mm in size (Kasai Sangyo Co., Ltd., Osaka, Japan). For the experimental groups, the test plates had the same size as their control counterparts, and were coated on one side with the test paint, as specified in the Performance Standards for Protection Coatings (PSPC) [34]. Surface treatment, measurement of Cu_2O concentration and the dynamic aging process of the plates were conducted according to the previous paper [33].

Table 1. Composition of the test paints used [33]. Adapted with permission from [33]. Copyright 2016 PLOS.

Composition/Paint Name	A-0	A-1	A-2	A-3	A-4	A-5
Cuprous oxide	0	5	10	20	30	40
Xylene	23	23.6	24	25	26	27
Methylisobutylketone	5	5	5	5	5	5
Base polymer	9	8.7	8.5	8	7.5	7
Rosin	9	8.7	8.5	8	7.5	7
Barium sulphate	50	45	40	30	20	10
Anhydrous ferric oxide	1	1	1	1	1	1
Oxidized polyethylene wax	1	1	1	1	1	1
Amide wax	2	2	2	2	2	2

Values in the table indicate mass %.

2.2. Laboratory Bioassay

2.2.1. Test Organism

Adults of *A.* (=*Balanus*) *amphitrite* were collected from Lake Hamana (Shizuoka, Japan: 34°41′ N, 137°35′ E). These were reared in the laboratory according to the method described in previous reports [18,19,23,35–39]. Adults of *A. amphitrite* were transferred into a tank filled with seawater adjusted to a temperature of 25 °C and were fed nauplii of the brine shrimp, *Artemia salina*. The seawater used was cartridge (1 μm) filtered and seawater diluted with purified water (Milli Q, Merck Millipore, Burlington, MA, USA) to adjust its salinity to 28 ± 0.5. The seawater was renewed daily to maintain the water quality in the rearing tank. To induce the release of nauplius larvae, adults were taken out of the tank and kept dry inside the incubator at 25 °C for at least 6 h. Nauplius larvae were released when adults were submerged back in the seawater adjusted to 25 °C. Newly hatched nauplius larvae exhibited positive phototactic behaviour and gathered around a light source. They were collected using a pipette and cultured according to the method described in a previous paper [19]. Briefly, clean two-liter glass beakers were used for culturing larvae of barnacles. For larval cultures of nauplii to the cyprid stage, 0.22 μm (Nylon filter membrane, Merck, Kenilworth, NJ, USA) of filtered seawater with the salinity adjusted to 28 using purified water was used. Streptomycin sulfate (FUJIFILM Wako Pure Chemical Corporation, Osaka, Japan) and Penicillin G sodium salt (FUJIFILM Wako Pure Chemical Corporation) were added into the culture water to final concentrations of 30 and 20 μg/mL, respectively. Cyprids were obtained after 5 days when cultured following the conditions presented in Table 2.

Table 2. Culture condition for nauplius larvae.

Condition	Remarks
Seawater	0.22 μm filtered seawater (Nylon filter membrane, Merck) with the salinity adjusted to 28 using purified water
Density of larvae	2 to 3 larvae per 1 mL
Diet and density	The diatom *Chaetoceros gracilis* (200,000 to 400,000 cells/mL), other diatoms can also be used, such as *Skeletonema costatum* (1,000,000 to 2,000,000 cells/mL)
Antibiotics	Streptomycin sulfate (30 μg/mL), Penicillin G sodium salt (20 μg/mL)
Water temperature	25 ± 1 °C
Light	The light intensity is approximately 3000 lux., and the photoperiod was 12 h. light, 12 h. dark period
Aeration	Approximately 20 mL/min.

Cyprids were collected in a beaker filled with 0.22 µm of filtered seawater and stored in a refrigerator at a temperature of 4 to 6 °C for about three days, before their use in bioassays. The condition of cyprids used in the bioassays for each culture batch was checked by observing mortality and settlement rates. That is, 10 individuals of 3-day-old cyprids were placed in each well of a 12-well polystyrene plate, with each well filled with 4 mL of the filtered seawater, and the number of juveniles that settled and dead cyprids were counted under a dissecting microscope after 48 h [40–42].

2.2.2. Evaluating the Inhibition Effects of Test Paints Using Cyprid Settlement as Index

The bioassay system consisted of the seawater storage tank (volume capacity of 20 L), peristaltic pump, polypropylene bioassay tank (diameter × depth = 80 mm × 55 mm, volume capacity = ca. 275), and reservoir tank (height × width × depth = 1500 mm × 900 mm × 450 mm) for waste seawater. The seawater storage tank and reservoir tank were placed outside the incubator. The schematic diagram of the system is shown in Figure 1. The size of the bioassay tank used was large enough to completely submerge the test plates. The bioassay tank also had a siphon tube made of glass for drainage. The bioassay tank was placed inside an incubator equipped with a temperature controller to keep temperature within 25 ± 1 °C. The set-up was illuminated for 12 h. each day with a light intensity of 3000 lux. During the bioassays, 0.22 µm filtered seawater was continuously charged into the bioassay tanks at a flow rate of 7 mL/min. Three rounds of bioassays were performed. Throughout the investigation, a control group with the control plates was always set and assessed simultaneously with the other test plates.

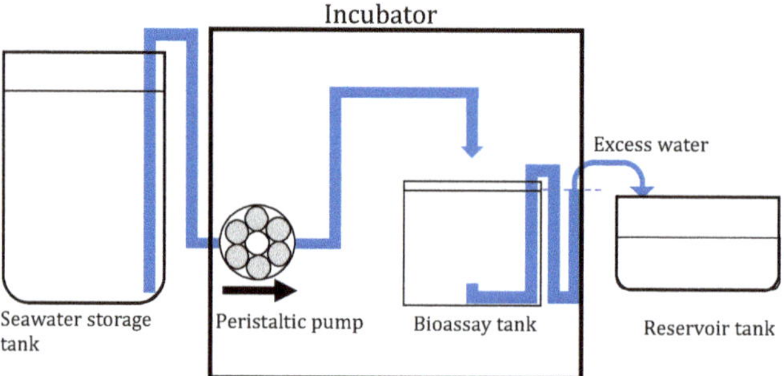

Figure 1. Schematic diagram of the bioassay system.

The bioassay was conducted using a triangular box, designed to enhance the settlement of cyprids by enclosing them in a small space within the three test plates. The box had one test/control plate and two white acrylic plates, assembled as shown in Figure 2. The triangular box was assembled so that the test or the control surface was positioned facing inward the triangular box. We found that larvae hardly settled on the surface of white acrylic plate (data not shown). Therefore, we used the white acrylic plates to promote cyprid settlement on the control and test surface.

Prior to the experiment with antifouling paints, a suitable control plate was determined using the present bioassay. Initially, the bioassay described above was used to determine a suitable control plate for inhibition experiments. Materials tested for the selection of a control plate were the white colored polystyrene (PS) plate, the grey colored PVC plate and the black colored PVC plate (Kasai Sangyo Co., Ltd., Osaka, Japan). Various treatments were also applied on the surface of the control plate by blasting with polishing agents, after the suitable material of the control plate was selected, and cyprid settlement on treated surfaces was investigated.

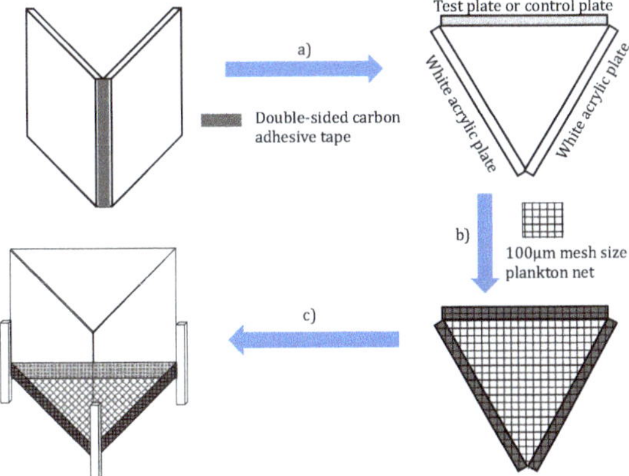

Figure 2. Assembling of the triangular box for the bioassay. (**a**) One test plate (or control plate) and two white acrylic plates were assembled using a carbon adhesive tape to form a triangular box; (**b**) the bottom-side of the triangular box was covered with a 100-μm mesh sized plankton net using a carbon adhesive tape; (**c**) plastic rods were attached to the triangular box to serve as legs for elevation (length of the rods from the bottom-side of the box: >10 mm).

The surfaces of the test and control plates were kept wet during the assembly of the triangular box. The bottom of the triangular box was covered with plankton net (mesh size NXX13, pore size: 100 μm). The test or control plate, two acrylic plates, and the plankton net were tightly adhered together using double-sided carbon tape (Nisshin EM Co., Ltd., Tokyo, Japan) as this material does not affect the swimming behavior of cyprids. Three plastic rods were attached to the triangular box to serve as legs, in order to elevate the bottom-side (net side) of the box by at least 10 mm from the bottom of the bioassay tank (Figure 3).

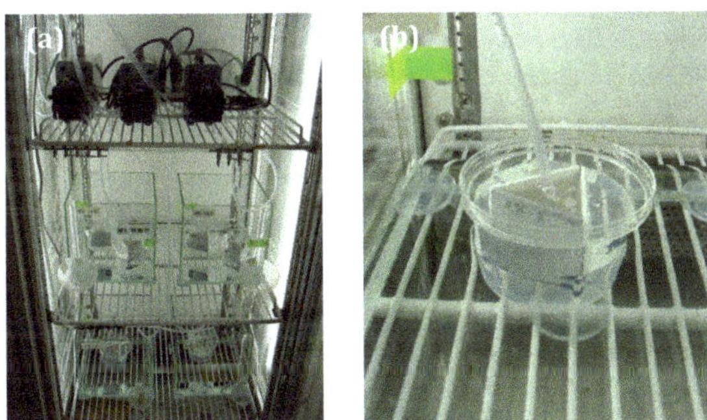

Figure 3. Photographs showing the bioassay system. (**a**) A photograph of peristaltic pumps and the triangular box inside the bioassay tanks placed in the incubator; (**b**) a photograph of the triangular box used in the bioassay.

The triangular boxes were assembled one day prior to the bioassay and then immersed overnight in the test seawater at room temperature. During this time, the seawater was stirred using a glass rod to prevent its stagnation near the triangular box. The triangular box was later taken out from the tank and immersed in 1 L of seawater for five minutes prior to the bioassay. The triangular box was then positioned inside the bioassay tank and the tank placed inside the incubator. The seawater was allowed to flow inside the bioassay tank from the topside of the triangular box. The flow rate was adjusted to prevent overflowing of seawater from the triangular box, as shown in Figure 4.

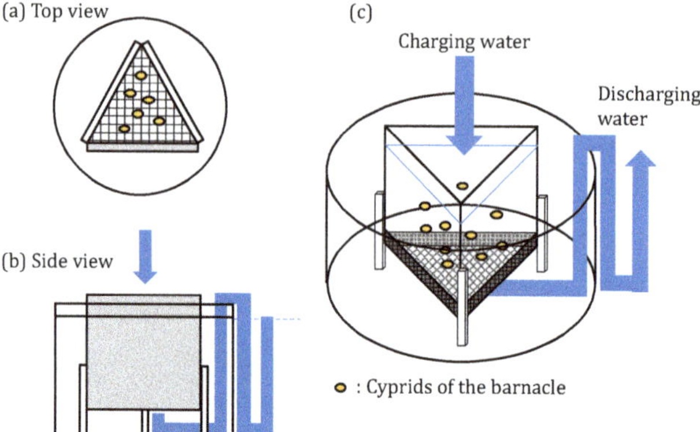

Figure 4. The triangular box deployed in the bioassay tank. (**a**) Top view of a triangular box containing cyprids deployed in the bioassay tank; (**b**) side view of a triangular box inside the bioassay tank; and (**c**) arrows indicating the direction of the flow of the test seawater being charged inside the triangular box and discharged from the bioassay tank.

One hundred cyprids were released inside the triangular box. Initial water temperature, salinity (CM-14P, DKK-TOA Corp., Tokyo, Japan), and pH (HM-30P, TKK-TOA Corp., Tokyo, Japan) of the test seawater in the bioassay tank were measured and monitored during the bioassay. Unattached cyprids and dead individuals inside the triangular box were immediately collected after measuring the water quality parameters at the end of the experiment. The triangular box was immediately disassembled after retrieving the cyprids and dead individuals. The number of settled juveniles, cyprids and dead individuals on each of the three surfaces of the triangular box and on the net were counted under a stereo microscope.

2.2.3. Verification of the Validity of the Bioassay

The number of settled juveniles on a multi-well plate, cyprids, and dead individuals were counted. The settlement ratio for verifying the validity of the bioassay (R_v) was calculated using Equation (1):

$$R_v = \frac{a}{a + b + c} \times 100 \tag{1}$$

where a indicates the number of juveniles settled on the surface of the multi-well plate; b, the number of cyprids (not settled); c, the number of dead individuals; R_v, the settlement ratio for verifying the validity of the bioassay (%).

The number of juveniles that settled on the test plate and the other surfaces (acrylic plates, edges of the test plate and net), the cyprids and the dead individuals were counted. The settlement ratio for test plate (R_t) was also calculated using Equation (2):

$$R_t = \frac{S_t}{S_t + a + b + c} \times 100 \tag{2}$$

where S_t indicates the number of settled juveniles on the surface of the test plate; a, the number of settled juveniles on the other surfaces; b, the number of cyprids (not settled); c, the number of dead individuals; R_t, the settlement ratio for test plate (%).

The settlement ratio for the control plate (R_c) was also calculated using Equation (3):

$$R_c = \frac{S_c}{S_c + a + b + c} \times 100 \tag{3}$$

where S_c indicates the number of juveniles that settled on the surface of the control plate; a, the number of juveniles that settled on the other surfaces; b, the number of cyprids (not settled); c, the number of dead individuals; R_c, the settlement ratio for the control plate (%).

The average values of the settlement ratio in all the experimental and control rounds were calculated using Equations (4) and (5), respectively:

$$A_t = \frac{\sum_{j=1}^{j}(R_{t1}^{j} + R_{t2}^{j} + \cdots R_{tn}^{j})}{\sum_{j=1}^{j} n^j} \tag{4}$$

where j indicates the run number; R_{tn}^{j}, the settlement ratio of the n-th test plate on the j-th run (%); n^j, the number of test plates on the j-th run; A_t, the average value of the settlement ratio in the experimental round (%).

$$A_c = \frac{\sum_{j=1}^{j}\left(R_{c1}^{j} + R_{c2}^{j} + \cdots R_{cn}^{j}\right)}{\sum_{j=1}^{j} n^j} \tag{5}$$

where j indicates the run number; R_{cn}^{j}, the settlement ratio of the n-th control plate on the j-th run (%); n^j, the number of control plates on the j-th run; A_c, the average value of the settlement ratio in the control round (%).

Finally, the relative settlement ratio of cyprids (R) was calculated using Equation (6):

$$R = \frac{A_t}{A_c} \tag{6}$$

2.3. Statistical Analysis

Statistical analysis, including one-way analysis of variance (ANOVA), nonparametric tests, and Holm-Sidak's multiple comparison test ($p < 0.05$) in the settlement assay, calculation of correlation coefficients, and variance analysis were performed using GraphPad Prism version 7.0 d for Mac OSX (GraphPad Software).

3. Results

3.1. Assessment of the Efficacy of Antifouling Paint on Cyprid Settlement Ratio

3.1.1. Parameters of Water Quality of the Test Water

The water temperature, salinity, and pH of the test water in the three laboratory experiments were controlled at 24.4 ± 1.2 °C, 27.4 ± 0.9, and 8.3 ± 0.1, respectively. The concentrations of Cu_2O in the test water of the control groups ranged from 2.2 to 2.5 µg/L after 48 h. Whereas, the concentrations of Cu_2O in the test water of the experimental groups were 2.3 µg/L (A-0), 2.9 µg/L (A-1), 13.3 µg/L (A-2), 4.7 µg/L (A-3), and 12.9 µg/L (A-4) after 48 h. The concentrations of Cu_2O in the test water of the A-5 groups ranged from 20.1 to 26.0 µg/L in the three experiments.

3.1.2. The Activity of Cyprids Used in the Bioassay

Settlement ratios ranged from 80% to 93%, and mortality was 0%. Therefore, cyprids used in the settlement assays were healthy and bioassays were verified as valid.

3.1.3. Relative Settlement Ratios (R) of Cyprids on Antifouling Paints

In the bioassay for the selection of a suitable material for the control plate, settlement ratio after the 48-h immersion period was calculated using Equation 5. Three replicates were conducted for each type of plate. Averages of the settlement ratios were 50.1% (SD = 16.5) for the black colored PVC, 28.8 % (SD = 11.9) for the grey colored PVC, and 25.6% (SD = 17.5) for the white colored PS plates, respectively.

The mortality of cyprids was less than 0.4% in each group. In another experiment, the surface of the black colored PVC was blasted with F-40, F-36 and F-20 polishing agents [43,44] respectively, and settlement ratios were also investigated after the 48-h immersion period. Four replicates were prepared for each blasted plate. Averages of the settlement ratios for F-20, F-36, and F-40 blasted plates were 53.2% (SD = 14.7), 24.9% (SD = 5.4), and 25.5% (SD = 9.7), respectively. Furthermore, the settlement ratio for the F-20 blasted black PVC was investigated at different times for validation, with four replicate plates tested at each time. Averages of the settlement ratios were 53.3% (SD = 14.7), 56.0% (SD = 10.7), and 44.9 % (SD = 14.5) for the three times. Time-related difference in settlement ratios was not recognized by variance analysis ($F = 1.0358$, $df = 3$, $p = 0.4181$). As a result, the black color PVC plate blasted with F-20 was used as a control plate in this bioassay.

Experiments with antifouling paints were conducted between October, November, and December of 2014, and in February of 2015. Each experiment was replicated three times; using a total of nine test plates for each experimental group and three test plates for the control group. To assess the inhibition effect of the test paints, statistical analysis was conducted by comparing results from the experimental groups with their respective controls. Results of the mortality and settlement bioassay are shown in Figures 5 and 6, respectively. Average mortality of cyprids in the control group was less than 1%, while mortalities in the 0, 5, 10, 20, and 30 wt.% Cu_2O experimental groups were less than 5%, and that in the 40 wt.% of Cu_2O group was less than 10%.

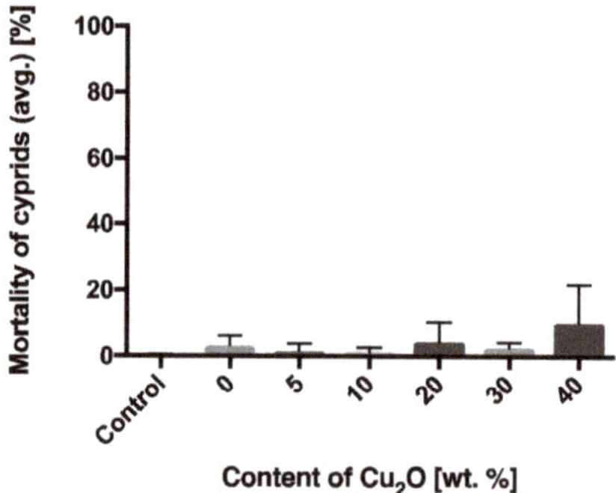

Figure 5. Average mortalities of cyprids in the control and experimental groups. Error bars indicate SDs.

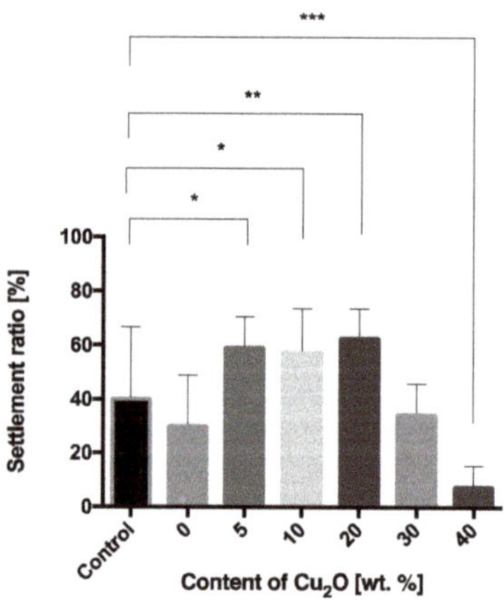

Figure 6. Settlement ratios in the control and experimental groups. Error bars indicate SDs. Pairwise comparisons of results between each experimental group and the control were conducted to establish significance of the difference (* $p < 0.05$, ** $p < 0.01$, and *** $p < 0.001$), where the symbols *, ** and *** corresponds to significant, very significant and extremely significant, respectively.

The settlement ratio in the control (A_c) was 40%, and settlement ratios in the 5 wt.% of Cu_2O, 10 wt.% of Cu_2O, and 20 wt.% of Cu_2O experimental groups (A_t) were higher than that of the control. The values of the settlement ratio in the 5 wt.% and 10 wt.% of Cu_2O experimental groups were significantly different from A_c ($p < 0.05$). In the 20 wt.% of Cu_2O group, the difference between A_t and A_c was very significant ($p < 0.01$), but there was no significant difference between A_c and A_t of the 30 wt.% of Cu_2O group. Concentrations below 30 wt.% of Cu_2O promoted settlement, whereas an inhibition of settlement was clearly observed at 40 wt.% of Cu_2O, where the difference in settlement ratio as compared to the control was extremely significant ($p < 0.0001$). To evaluate the settlement inhibition effect of the test paints, data were normalized by calculating the relative settlement ratios of cyprids (R) in the experimental groups with respect to their respective controls.

The R values of the paints containing different concentrations of Cu_2O are shown in Figure 7. R values showed a non-linear relationship with the Cu_2O content in the paint ($r^2 = 0.3403$, non-linear curve fitting, second-order polynomial (quadratic equation)), where R increased with increasing Cu_2O content of the paint, at up to 20 wt.%, but decreased thereafter. The results also showed that 50% settlement inhibition (EC_{50}) was obtained at approximately 38.0 wt.% of Cu_2O through interpolation of the fitting curve. Moreover, the R value at 40 wt.% of Cu_2O was less than 0.5 and the relative settlement ratio in this group obviously differed from that in the other experimental groups. Therefore, R values of 0.5 and lower can be considered as inhibition settlement ratio.

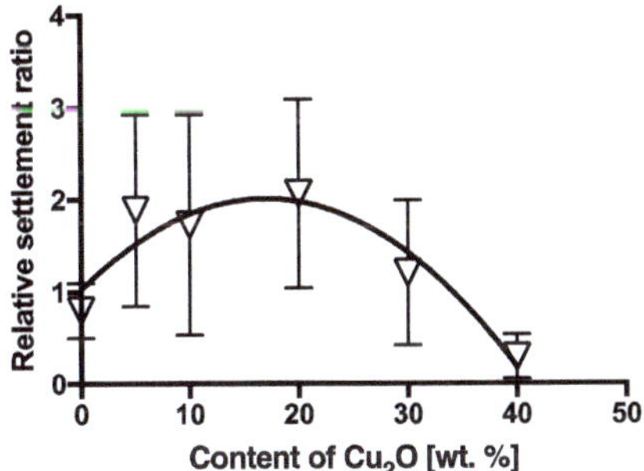

Figure 7. Relative settlement ratios on paints containing different concentrations of Cu_2O. Error bars on the closed triangles indicate SDs of the relative settlement ratios of cyprids.

4. Discussion

4.1. The Concept of the Flow-Through Bioassay Designed for Barnacles in the Laboratory

Biocides have been primarily screened in-vitro for antifouling activity and toxicity using multi-well plates [2]. Even experiments investigating the tolerance of nauplius larvae to copper stress have been designed in a still water condition [17,45]. It is evident that literature on the evaluation of the efficacy of antifouling agents mostly described bioassay conducted in a still water condition.

Under such a condition, the leached biocides accumulate in the experimental system as the experimental period progresses. This increase in the concentration of biocides may affect the physiological condition of the test organisms. As a result, it is difficult to appropriately conduct an experimental bioassay. To address this difficult situation, the authors designed an experimental method with a flow-through system that renewed the test water continuously inside the experimental system. This system followed the same concept of the flow-through bioassay designed for the mussel *M. galloprovincialis* [33]. It is unique compared to other flow-through bioassay systems previously reported (e.g., [15]) due to the following features: (a) it has a low (ca. ~0.42 L/h) flow rate and is compact (vessel volume ca. 275 mL); (b) a constant density of cyprids inside the triangular vessel can be adjusted and ensured; and (c) enclosing cyprids inside a triangular vessel and using inert white acrylic plates enhanced cyprid settlement on the surface of test plates, depending on the efficacy of the antifouling coating on the test plate. The barnacle *Amphibalanus amphitrite* was selected as the test organism because it is reported as one of the major macrofoulers [46,47]. However, no literature, except for this study, has ever reported on a barnacle settlement assay that is compact and that simulated actual conditions of the painted surface of ship hulls. Moreover, no study prior to this has introduced a barnacle settlement assay with a flow-through system in a small vessel, thereby using only a small volume of seawater.

The material chosen for the control plate was black PVC because it made observation of settled juveniles easier. The surface of the control plates was blasted with the polishing agent, F-20. This combination of material and surface treatment of the control plates resulted in a higher settlement ratio in the control group.

4.2. The Repellent Effect of Cu_2O on Barnacles

In a toxicity study conducted under a still water condition, the mortality of barnacles from $CuSO_4 \cdot 5H_2O$, which was used as a positive control, was 10% at 1000 µg/L; and mortality was 50% at 3000 µg/L [25]. The repellent effect on barnacles was detected from 100 µg/L with 20% inhibition of settlement [25]. The EC_{50} value was 300 µg/L, and 90% inhibition was at 1000 µg/L [25]. At the end of the experiment in this study, the Cu_2O concentration of the test water was 24 µg/L in the group coated with the paint A-5 containing 40 wt.% of Cu_2O. This concentration was almost 1/10 to 1/40 times lower than the value in the abovementioned still water bioassay [25]. Mortality of cyprids in the paint A-5 group was 10% after 48 h.

In this study, the A_t values of 5, 10, and 20 wt.% of Cu_2O groups were higher than that of the control. Hence, the concentration below 30 wt.% of Cu_2O promoted cyprid settlement. The result indicates that tolerance to copper stress was below 20 wt.% of Cu_2O content in the test paints. An indication of the relative tolerance of various organisms to toxic paints was reported previously [48]. *Amphybalanus amphitrite* was reported to be more tolerant to Cu than other foulers. The total number of adult barnacles attached to the toxic paint surface was less than 1% of those growing on the non-toxic control. *Amphybalanus amphitrite* comprised more than 90% of the total adult barnacle population on the toxic surface, although it comprised only 7.5% of the population attached to the control surface [48]. Regarding the toxic effect of copper on the larval development of barnacles, larvae in the advanced developmental stages may have developed better physiological mechanisms to regulate the uptake or increase the excretion of the toxicant [45]. As part of an antifouling investigation, the survival time and O_2 consumption of adult *A. amphitrite* exposed to different Cu concentrations were investigated [49]. The result showed that O_2 consumption rate of barnacles decreased during respiration, and data on survival time indicated that *A. amphitrite* was more tolerant to Cu than *B. tintinnabulum*. It was argued that tolerance to Cu was derived from metallothioneins [49]. These proteins were found to increase with exposure to metal concentrations, leading to the sequestration and detoxification of metals to some extent [49]. The concentration of copper in barnacles from uncontaminated sites was reported [50,51], and it demonstrated that the barnacle has the potential to accumulate high concentrations of copper and showed strong net accumulation of copper. In this situation, all incoming copper was accumulated for detoxification [52], and barnacles from Cu-contaminated sites had many type B (more heterogenous in shape and always containing sulphur in association with metals that include copper and zinc [53,54]) Cu-rich granules, probably resulting from lysosomal breakdown of metallothionein binding copper [51].

Another possible explanation of this phenomenon is the effect as an inducer. It has been demonstrated that chemical compounds such as epinephrine, phenylephrine, clonidine, KCl, NH_4Cl, and organic solvent induced mussel larval metamorphosis [55]. The authors explained that sub-toxic levels of these compounds could have triggered larval metamorphosis by physiologically "shocking" the larvae since dead larvae were observed in concentrations that induced metamorphosis [55]. In addition, the enhancement of the settlement of *Capitella* sp. I larvae by H_2S had a sub-lethal effect [56]. It was also hypothesized that sub-lethal concentrations of H_2S could trigger larval settlement and/or metamorphosis by physiologically "shocking" the larvae [56]. In this report, the authors explained that the stimulus for settlement was a chemical cue associated with chemical substances to some extent. Copper was essential for crustaceans and mollusks, and excess amounts of chemical substances were toxic leading to the inhibition of the settlement of barnacles [52]. Such an increase in the settlement ratio of barnacles to Cu_2O content that was between 5 and 20 wt.% occurred in the present study.

In order to validate the antifouling efficacy of the test paints, results of laboratory bioassays and field experiments were compared. The reason for this comparison is because data of biofouling obtained from raft and patch experiments are affected by geological and seasonal variations and assessing the performance of antifouling paints in these experiments takes time [57,58]. Details of the field experiments were reported by the authors in a previous paper [33]. The immersion period during the field experiments was 28 days, after which the degree of fouling on the test plates was evaluated. Figure 8 shows the relationship between R and ranks of fouling. Both the R and the ranks of fouling showed a similar decreasing tendency with increasing Cu_2O content at more than 20 wt.% at St.1 and St.2 (St.1: r^2 = 0.9759, non-linear curve fitting, one-phase decay; St.2: r^2 = 0.9135, non-linear curve fitting, one-phase decay [33]). As a result, the proposed method using barnacles can sufficiently verify the repellant effect of test paints within 48 h. This method can also be used on other fouling species, such as algae, in future studies prior to the field experiment [59].

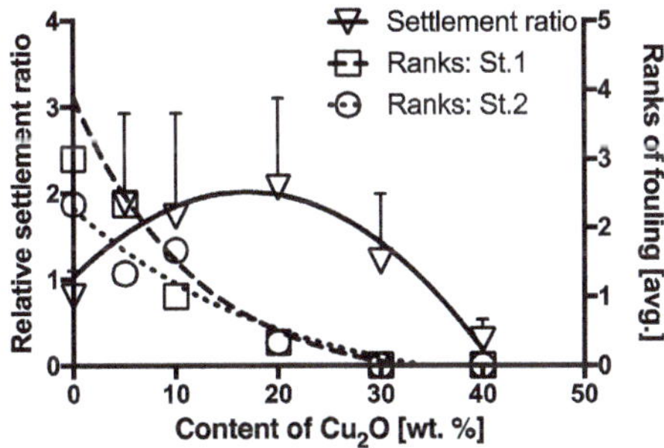

Figure 8. The relative settlement ratio of barnacle larvae and the degree of fouling in the field experiment on relation to the concentration of Cu_2O on the test paints. The relative settlement ratios from bioassay experiments ($\triangledown$) and the degrees of fouling at St. 1 ($\square$) and St.2 ($\bigcirc$) cited from a previous paper [32] were plotted in relation to the concentration of Cu_2O in the paints. Error bars on the open triangles indicate SDs.

5. Conclusions

The newly proposed method for evaluating the efficacy of biocide-releasing antifouling paints in a flow-through system using *A. amphitrite* was validated. A reproduceable and effective laboratory bioassay was established by evaluating the antifouling efficacy of test paints that were prepared with varying Cu_2O contents. To simulate the actual condition of ship hulls, dynamic aging of test plates was conducted. A positive correlation between the Cu_2O content and the repellant effect of the paint on barnacle larvae was observed at concentrations of more than 30 wt.%. Comparison of the results between laboratory bioassays using barnacle larvae and of field experiments revealed a highly consistent relationship between the two. Our results support the use of barnacles in evaluating the efficacy of antifouling paints because this bioassay does not test toxicity and is precise in that it imitates the exposed condition of paint on ship hulls, when the ship is in a stationary state after voyage. The novelty of this method lies in the aspect of assessing antifouling efficacy of paints by the evaluation of the behavior of barnacles inside the triangular box in a flow-through system, and so far, this is the only study that evaluates barnacle larval behavior to antifouling paints in situ in a flow-through system. This study also proved to be a significantly consistent method for assessing the effectiveness of present or future antifouling paints.

Author Contributions: Conceptualization, S.K., C.G.P.S., K.M., and I.K.; Data Curation, R.K., S.K., C.G.P.S., Y.S., and K.M.; Formal Analysis, R.K. and S.K.; Funding Acquisition, R.K. and H.A.; Investigation, R.K., S.K., C.G.P.S., Y.S., and K.M.; Methodology, R.K., S.K., C.G.P.S., K.M., and I.K.; Project Administration, R.K. and H.A.; Resources, H.A.; Supervision, H.A. and I.K.; Validation, S.K., C.G.P.S., and K.M.; Visualization, I.K.; Writing—Original Draft, R.K.; Writing—Review and Editing, S.K., C.G.P.S., and K.M.

Funding: This research was supported by the research project on "The study on management of biofouling (2012)" of the Japan Ship Technology Research Association (JSTRA) (project ID: 20011965347, continued project ID: 2000070441C99), which was funded by the Nippon Foundation (http://www.nippon-foundation.or.jp).

Acknowledgments: The authors would like to express their appreciation to researchers Hiroshi Masuda, Hirohisa Mieno, Kazuki Kouzai of Chugoku Marine Paints Co., Ltd., Mamoru Shimada of Nippon Paint Marine Coatings Co., Ltd., Eiichi Yoshikawa of ME consulting, and Tetsuya Senda of JSTRA for their useful comments and contribution to the study. Finally, the authors wish to acknowledge A.S. Clare of Newcastle University for his editing and constructive suggestions on the early version of this manuscript.

Conflicts of Interest: The authors have declared that no competing interests exist. The funders had no role in the design of the study; in the collection, analyses, or interpretation of data; in the writing of the manuscript, and in the decision to publish the results.

References

1. Richmond, M.D.; Seed, R. A review of marine macrofouling communities with special reference to animal fouling. *Biofouling* **1991**, *3*, 151–168. [CrossRef]
2. Briand, J.F. Marine antifouling laboratory bioassays: An overview of their diversity. *Biofouling* **2009**, *25*, 297–311. [CrossRef] [PubMed]
3. Carlton, J.T. Pattern, process, and prediction in marine invasion ecology. *Biol. Conserv.* **1996**, *78*, 97–106. [CrossRef]
4. Coutts, A.D.M.; Taylor, M.D. A preliminary investigation of biosecurity risks associated with biofouling on merchant vessels in New Zealand. *N. Z. J. Mar. Freshw. Res.* **2004**, *38*, 215–229. [CrossRef]
5. Schultz, M.P. Effects of coating roughness and biofouling on ship resistance and powering. *Biofouling* **2007**, *23*, 331–341. [CrossRef] [PubMed]
6. Otani, M.; Oumi, T.; Uwai, S.; Hanyuda, T.; Prabowo, R.E.; Yamaguchi, T.; Kawai, H. Occurrence and diversity of barnacles on international ships visiting Osaka Bay, Japan, and the risk of their introduction. *Biofouling* **2007**, *23*, 277–286. [CrossRef] [PubMed]
7. Davidson, I.C.; Brown, C.W.; Sytsma, M.D.; Ruiz, G.M. The role of containerships as transfer mechanisms of marine biofouling species. *Biofouling* **2009**, *25*, 645–655. [CrossRef] [PubMed]
8. Piola, R.F.; Dafforn, K.A.; Johnston, E.L. The influence of antifouling practices on marine invasions. *Biofouling* **2009**, *25*, 633–644. [CrossRef]
9. Coutts, A.D.M.; Valentine, J.P.; Edgar, G.J.; Davey, A.; Wilson, B.B. Removing vessels from the water for biofouling treatment has the potential to introduce mobile non-indigenous marine species. *Mar. Pollut. Bull.* **2010**, *60*, 1533–1540. [CrossRef]
10. Coutts, A.D.M.; Piola, R.; Hewitt, C.L.; Connell, S.D.; Gardner, J.P.A. Effects of vessel voyage speed on survival of biofouling organisms: Implications for translocation of non-indigenous marine species. *Biofouling* **2010**, *26*, 1–13. [CrossRef]
11. Hopkins, G.A.; Forrest, B.M. A preliminary assessment of biofouling and non-indigenous marine species associated with commercial slow-moving vessels arriving in New Zealand. *Biofouling* **2010**, *26*, 613–621. [CrossRef] [PubMed]
12. Campbell, M.L.; Hewitt, C.L. Assessing the port to port risk of vessel movements vectoring non-indigenous marine species within and across domestic Australian borders. *Biofouling* **2011**, *27*, 631–644. [CrossRef] [PubMed]
13. Clare, A.S.; Matsumura, K. Nature and perception of barnacle settlement pheromones. *Biofouling* **2000**, *15*, 57–71. [CrossRef] [PubMed]
14. Rittschof, D.; Cohen, J.H. Crustacean peptide and peptide-like pheromones and kairomones. *Peptides* **2004**, *25*, 1503–1516. [CrossRef] [PubMed]

15. Pansch, C.; Jonsson, P.R.; Berglin, M.; Pinori, E.; Wrange, A. A new flow-through bioassay for testing low-emission antifouling coatings. *Biofouling* **2017**, *8*, 613–623. [CrossRef] [PubMed]

16. Rittschof, D.; Bransconb, E.S.; Castlow, J.D. Settlement and behavior in relation to flow and surface in larval barnacle, *Balanus amphitrite* Darwin. *J. Exp. Mar. Biol. Ecol.* **1984**, *82*, 131–146. [CrossRef]

17. Rittschof, D.; Clare, A.S.; Gerhart, D.J.; Mary Sister, A.; Bonaventura, J. Barnacle in vitro assays for biologically active substances: Toxicity and settlement assays using mass cultured *Balanus amphitrite amphitrite* darwin. *Biofouling* **1992**, *6*, 115–122. [CrossRef]

18. Kitamura, H. Methods of larval culture and water-flow assay for larval settlement of the barnacle, *Balanus amphitrite*. *Sess. Org.* **1999**, *15*, 15–21. [CrossRef]

19. Yoshimura, E.; Nogata, Y.; Sakaguchi, I. Simple methods for mass culture of barnacle larvae. *Sess. Org.* **2006**, *23*, 39–42. [CrossRef]

20. Aldred, N.; Scardino, A.; Cavaco, A.; Nys, R.D.; Clare, A.S. Attachment strength is a key factor in the selection of surfaces by barnacle cyprids (Balanus Amphitrite) during settlement. *Biofouling* **2010**, *26*, 287–299. [CrossRef]

21. Chen, H.N.; Tdang, L.M.; Chong, V.C.; Chan, B.K.K. Worldwide genetic differentiation in the common fouling barnacle, *Amphibalanus amphitrite*. *Biofouling* **2014**, *30*, 1067–1078. [CrossRef] [PubMed]

22. Kon-ya, K.; Miki, W. All-seasonal assay for antifouling substances using reared barnacle larvae. *J. Mar. Biotechnol.* **1994**, *1*, 193–195.

23. Kon-ya, K.; Miki, W. Effect of environmental factors on larval settlement of the barnacle *Balanus amphitrite* reared in the laboratory. *Fish. Sci.* **1994**, *60*, 563–565. [CrossRef]

24. Usami, M. Prevention of barnacle settlement by electro-conductive coating system. *Sess. Org.* **1998**, *15*, 1–4. [CrossRef]

25. Kitano, Y.; Yokoyama, A.; Nogata, Y.; Shishima, K.; Yoshimura, E.; Chiba, K.; Tada, M.; Sakaguchi, I. Synthesis and anti-barnacle activities of novel 3-Isocyanotheonellin analogues. *Biofouling* **2003**, *19*, 187–192. [CrossRef] [PubMed]

26. Pettitt, M.E.; Henry, S.L.; Callow, M.E.; Callow, J.A.; Clare, A.S. Activity of commercial enzymes on settlement and adhesion of cypris larvae of the barnacle *Balanus amphitrite*, spores of the green alga *Ulva linza*, and the diatom *Navicula perminuta*. *Biofouling* **2004**, *20*, 299–311. [CrossRef] [PubMed]

27. Horiuchi, R.; Kobayashi, S.; Kameyama, Y.; Mizutani, M.; Komotori, J.; Katsuyama, I. Barnacle settlement behavior on controlled micro-textured surfaces. *Zair. Kankyo* **2009**, *58*, 302–307. (In Japanese) [CrossRef]

28. Murosaki, T.; Noguchi, T.; Kakugo, A.; Putra, A.; Kurokawa, T.; Furukawa, H.; Osada, Y.; Gong, J.P.; Nogata, Y.; Matsumura, K.; et al. Antifouling activity of synthetic polymer gels against cyprids of the barnacle (*Balanus amphitrite*) in vitro. *Biofouling* **2009**, *25*, 313–320. [CrossRef]

29. Petrone, L.; Fino, A.D.; Aldred, N.; Sukkaew, P.; Ederth, T.; Clare, A.S.; Liedberg, B. Effects of surface charge and Gibbs surface energy on the settlement behavior of barnacle cyprids (*Balanus amphitrite*). *Biofouling* **2011**, *27*, 1043–1055. [CrossRef]

30. Guo, S.; Lee, H.P.; Teo, S.L.M.; Khoo, B.C. Inhibition of barnacle cyprid settlement using low frequency and intensity ultrasound. *Biofouling* **2012**, *28*, 131–141. [CrossRef]

31. Maleschlijski, S.; Bauer, S.; Fino, A.D.; Sendra, G.H.; Clare, A.S.; Rosenhahn, A. Barnacle cyprid motility and distribution in the water column as an indicator of the settlement-inhibiting potential of nontoxic antifouling chemistries. *Biofouling* **2014**, *30*, 1055–1065. [CrossRef] [PubMed]

32. Katsuyama, I.; Kado, R.; Kominami, H.; Kitamura, H. A screening method for the test substances on attachment using larval barnacle, *Balanus amphitrite*, in the laboratory. *Mar. Fouling* **1992**, *9*, 13–14. [CrossRef]

33. Kojima, R.; Kobayashi, S.; Satuito, C.G.P.; Katsuyama, I.; Ando, H.; Seki, Y.; Senda, T. A method for evaluating the efficacy of antifouling paints using *Mytilus galloprovincialis* in the laboratory in a flow-through system. *PLoS ONE* **2016**, *11*, e0168172. [CrossRef] [PubMed]

34. IMO. Resolution MSC 215 (82): Performance Standards for Protection Coatings for Dedicated Seawater Ballast Tanks in All Types of Ships and Double-Side Skin Spaces of Bulk Carriers. 82/24/Add.1, ANNEX 1; IMO: London, UK, 2006.

35. Kado, R.; Hirano, R. Rearing methods of planktonic larvae of marine sessile animals. *Mar. Fouling* **1979**, *1*, 11–19. [CrossRef]

36. Kitamurta, H.; Nakashima, Y. Influence of storage temperatures and period on settlement rate and substrate discrimination in cyprids of the barnacle *Balanus amphitrite*. *Fish. Sci.* **1996**, *62*, 998–999. [CrossRef]

37. Sakaguchi, I. Bivavle. In *Attaching Organisms and Aquaculture*; Kajihara, T., Ed.; Kouseishya Kouseikaku Press: Tokyo, Japan, 2007; Chapter 9; pp. 100–107. (In Japanese)

38. Tomoda, K.; Ninomiya, K.; Azuma, K.; Suzuki, T.; Satuito, C.G.; Kitamura, H. Larval settlement of the barnacle *Amphibalanus amphitrite* on coal-flyash concrete plates in the laboratory. *Sess. Org.* **2008**, *25*, 25–29. [CrossRef]

39. Nogata, Y.; Tokikuni, N.; Yoshimura, E.; Sato, K.; Endo, N.; Matsumura, K.; Sugita, H. Salinity limitation on larval settlement of four barnacle species. *Sess. Org.* **2011**, *28*, 47–54. [CrossRef]

40. Kawahara, H.; Tamura, R.; Ajioka, S.; Shizuri, Y. Convenient assay for settlement inducing substances of barnacles. *Mar. Biotechnol.* **1999**, *1*, 98–101. [CrossRef]

41. Dahms, H.U.; Jin, T.; Qian, P.T. Adrenoceptor compounds prevent the settlement of marine invertebrate larvae: *Balanus amphitrite* (Cirripedia), *Bugula neritina* (Bryozoa), and *Hydroides elegans* (Plychaeta). *Biofouling* **2004**, *20*, 313–321. [CrossRef]

42. Zhou, X.; Xu, Y.; Jin, C.; Qian, P.Y. Reversible anti-settlement activity against *Amphibalanus* (=*Balanus*) *amphitrite*, *Bulgula neritina*, and *Hydroides elegans* by a nontoxic pharamaceutical compounds, mizolastine. *Biofouling* **2009**, *25*, 739–747. [CrossRef]

43. ISO. *ISO 8486-1:1996 Bonded Abrasives—Determination and Designation of Grain Size Distribution—Part 1: Macrogrits F4 to F220*; ISO: Geneva, Switzerland, 1996.

44. Japanese Standard Association. *Bonded Abrasives—Determination and Designation of Grain Size Distribution—Part 1: Macrogrits F4 to F220*; JIS R 6001-1; Japanese Standard Association: Tokyo, Japan, 2017.

45. Qiu, J.W.; Thiyagarajan, V.; Cheung, S.; Qian, P.Y. Toxic effects of copper on larval development of the barnacle *Balanus amphitrite*. *Mar. Pollut. Bull.* **2005**, *51*, 688–693. [CrossRef] [PubMed]

46. Khandepaker, L.; Anil, A.C. Underwater adhesion: The barnacle way. *Int. J. Adhes. Adhes.* **2007**, *27*, 165–172. [CrossRef]

47. Aldred, N.; Clare, A.S. The adhesion strategies of cyprids and development of barnacle-resistant marine coatings. *Biofouling* **2008**, *24*, 351–363. [CrossRef] [PubMed]

48. Weiss, C.M. The comparative tolerance of some fouling organisms to copper and mercury. *Biol. Bull.* **1947**, *93*, 56–63. [CrossRef]

49. Rao, Y.P.; Devi, V.U.; Rao, D.G.V.P. Copper toxicity in tropical barnacles, *Balanus amphitrite* and *Balanus tintinnabulum tintinnabulum*. *Water Air Soil Pollut.* **1985**, *27*, 109–115. [CrossRef]

50. Rainbow, P.S.; Blackmore, G.; Wang, W.X. Effects of previous field-exposure history on the uptake of trace metals from water and food by the barnacle *Balanus amphitrite*. *Mar. Ecol. Prog. Ser.* **2003**, *259*, 201–213. [CrossRef]

51. Rainbow, P.S. Trace metal bioaccumulation: Models, metabolic availability and toxicity. *Environ. Int.* **2007**, *33*, 576–582. [CrossRef]

52. Rainbow, P.S. Phylogeny of trace metal accumulation in crustaceans. In *Metal Metabolism in Aquatic Environment*; Langston, W.J., Bebianno, M.J., Eds.; Chapman & Hall: London, UK, 1998; pp. 285–319.

53. Walker, G. Copper granules in the barnacle *Balanus balanoides*. *Mar. Biol.* **1997**, *39*, 343–349. [CrossRef]

54. Rainbow, P.S. Heavy metals in barnacles. In *Barnacle Biology*; Southward, A.J., Ed.; A. A. Balkema: Rotterdam, The Netherlands, 1987; pp. 405–417.

55. Yang, J.L.; Satuito, C.G.; Bao, W.Y.; Kitamura, H. Induction of metamorphosis of pediveliger larvae of the mussel *Mytilus galloprovincialis* Lamarck, 1819 using neuroactive compounds, KCl, NH_4Cl and organic solvents. *Biofouling* **2008**, *24*, 461–470. [CrossRef] [PubMed]

56. Dubilier, N. H_2S-A settlement cure or a toxic substance for Cplitella sp. I Larvae? *Biol. Bull.* **1998**, *174*, 30–38. [CrossRef]

57. Ketchum, B.H.; Ferry, J.D.; Redfield, A.C. Evaluation of antifouling paints by leaching rate determinations. *Ind. Eng. Chem.* **1945**, *37*, 456–460. [CrossRef]

58. Berntsson, K.M.; Jonson, P.R. Temporal and spatial patterns in recruitment and succession of a temperate marine fouling assemblage: A comparison of static panels and boat hulls during the boating season. *Biofouling* **2003**, *19*, 187–195. [CrossRef] [PubMed]
59. Gatley-Montross, C.M.; Finlay, J.A.; Aldred, N.; Cassady, H.; Destino, J.F.; Orihuela, B.; Hickner, M.A.; Clare, A.S.; Rittschof, D.; Holm, E.R.; et al. Multivariate analysis of attachment of biofouling organisms in response to material surface characteristics. *Biointerphases* **2017**, *12*, 051003. [CrossRef] [PubMed]

Article

Silver Nanoparticle-Based Paper Packaging to Combat Black Anther Disease in Orchid Flowers

Bang-on Nokkrut, Sawitree Pisuttipiched, Somwang Khantayanuwong and Buapan Puangsin *

Department of Forest Products, Faculty of Forestry, Kasetsart University, Bangkok 10900, Thailand;
Puziz_Model@hotmail.co.th (B.N.); fforsap@ku.ac.th (S.P.); fforsok@ku.ac.th (S.K.)
* Correspondence: fforbpp@ku.ac.th; Tel.: +66-2942-8109 (ext. 1918)

Received: 15 December 2018; Accepted: 11 January 2019; Published: 14 January 2019

Abstract: Metal nanoparticles have been reported to have a high antimicrobial activity against fungi, bacteria, and yeasts. In this study, we aimed to synthesize silver nanoparticles (AgNPs) using a chemical reduction method at 90 °C. The obtained AgNPs were used as an antifungal coating on packaging paper, to control the growth of *Colletotrichum gloeosporioides* in cut orchid flowers during the shipping process. The AgNPs were characterized by a UV-Vis spectroscopy and atomic force microscope (AFM). The results indicated that their shape was spherical and homogenous, with an average size of 47 nm. An AgNPs concentration of 20 and 50 particles per million (ppm), mixed with starch, was prepared as the coating solution. The paper coated with a concentration of 50 ppm exhibited a significant antifungal activity against *C. gloeosporioides* compared to 20 ppm. The coated paper had a higher water resistance and better mechanical properties compared to the uncoated paper. Additionally, we observed a significant reduction in the number of orchid inflorescence anthers, infected by *C. gloeosporioides*, when stored in the coated boxes. The current study demonstrates that paper boxes, coated with AgNPs, can be used in controlling the *C. gloeosporioides* infection during storage of cut orchid flowers.

Keywords: black anther disease; orchid cut flower; silver nanoparticles

---

1. Introduction

The orchid is one of the most important commercially viable ornamental plants in Thailand, especially the cut flowers and potted plants [1,2]. Around 1300 species and 180–190 orchid genera have been reported, that are grown widely across the country [3]. Thailand is one of the world's largest exporters of cut orchid flowers and has had a long history of orchid trading around the world [2]. The total orchid flower export business is valued at around 59–70 million US$ (approximately 1949–2307 million Thai Baht) [2]. In addition, the exports of cut orchid flowers experienced a decrease of about 24.2 million tons, valued at approximately 67.5 million US$ (approximately 2228 million Thai Baht), in 2017. Black anther, caused by the phytopathogen *Colletotrichum gloeosporioides*, causes a significant reduction in the postharvest quality of cut orchid flowers, especially during the rainy season [3,4]. The symptoms of this disease include a black spot on the anther of the flower [5], lowering the quality and a shorter vase life, and in turn, a reduction in the export value.

Synthesized fungicides, such as thiabendazole, prochloraz, azoxystrobin, and chlorothalonil, have been commonly used to control the growth of *C. gloeosporioides* in the orchid, during postharvest shelf life; however, their excessive use has resulted in the fungi becoming increasingly resistant to the fungicides. At the same time, there has been an increasing concern related to consumer safety [5,6]. The development of an antimicrobial packaging paper, with properties that can prolong the shelf life of the product during storage or transportation, while maintaining an acceptable quality, has been gaining the attention of researchers [7,8]. Different types of antimicrobial agents, such as silver nanoparticles

(AgNPs) [6,9,10], zinc pyrithione [11], benzimidazole [12], organic acids [13], borate [14], and plant extraction [15], have been reported to be potential coating material in paper boxes, in order to control the growth of *C. gloeosporioides*. However, organic and natural biological antimicrobials have been reported to be less stable at higher temperatures and have high volatility, compared to inorganic ones [16,17], which may result in a limited application.

To our knowledge, studies related to the use of antimicrobial coating on paper boxes for the inhibition of *C. gloeosporioides* in harvested orchid flower packaging in Thailand have yet to be made. Therefore, the main objective of this study was to synthesize AgNPs by a chemical reduction method and use it as an antifungal agent. The packaging paper was coated with an appropriate amount of AgNPs, mixed with a starch solution, to increase its antifungal properties. The morphology, basis weight, thickness, mechanical properties, and water resistance of the paper were evaluated. Additionally, the antifungal activity of coated paper, to combat *C. gloeosporioides*, by disc diffusion, was evaluated. Finally, the efficacy of the antifungal coating, in inhibiting the proliferation of black anther disease in stored cut orchid flowers, was also investigated.

2. Materials and Methods

2.1. Materials

Uncoated paper (134 g/m^2), commonly used for the storage of orchid flowers, was obtained from Mahachai Kraft Paper Co. Ltd. (Samutsakorn, Thailand). Hydrophobic starch (FILMKOTE 370TM) was supplied by National Starch and Chemical (Thailand) Co. Ltd. (Samutprakan, Thailand). Chemicals used to prepare AgNPs, i.e., Silver nitrate (AgNO$_3$) and Sodium hydroxide (NaOH), were purchased from Merck Co. (Darmstadt, Germany). Trisodium citrate (Na$_3$C$_6$H$_5$O$_7$) and sodium borohydride (NaBH$_4$) were purchased from Ajax Finechem Co. (Victoria, Australia). Potato dextrose agar (PDA) was purchased from HiMedia Laboratories Pvt. Ltd. (Mumbai, India). *C. gloeosporioides* was obtained from the Department of Agriculture, Ministry of Agriculture and Cooperatives (Bangkok, Thailand). The cut orchid flowers (*Dendrobium sonia*) were collected from the Siamtaiyoo farm Co. Ltd. (Samut Sakhon, Thailand).

2.2. Synthesis of Silver Nanoparticles

AgNPs were prepared by the chemical synthesis method outlined in Agnihotri et al. [18]. Briefly, a mixture of aqueous solution containing 24 mL of sodium borohydride (2×10^{-3} mol·dm^{-3}) and 24 mL of trisodium citrate (4.28×10^{-3} mol·dm^{-3}) was heated to 60 °C for 30 min, in the absence of light. Silver nitrate (1×10^{-3} mol·dm^{-3}) solution (2 mL) was then added to the mixture, and heated from 60 to 90 °C. The reaction was allowed to continue for an additional 20 min and then cooled down to room temperature. The UV–visible spectrum properties of the synthesized AgNPs solution were determined, using a spectrophotometer (UV-1800, Shimadzu Corp., Kyoto, Japan), at wavelengths ranging between 300–700 nm. The dimensions of AgNPs were examined through an atomic force microscope (MFP-3D (Bio), Asylum Research Corp., Santa Barbara, CA, USA). A drop of AgNP solution was placed on a glass plate and dried in a desiccator for 24 h, prior to an analysis in the atomic force microscope (AFM), set on the tapping mode and a scan rate of 0.80 Hz.

2.3. Preparation of Antifungal Coating Solution

The antifungal coating solution was prepared using 8 g of hydrophobic starch, added to 100 mL of deionized water. The mixture was then heated and stirred at 90 ± 3 °C for 30 min. The obtained starch solution was cooled down to 65 ± 3 °C and the AgNP solution added to the predetermined quantity (0, 20, and 50 ppm) before use.

2.4. Preparations of Antifungal Coating Papers

The blended solution was coated on multiple papers (180 × 180 mm) using the bar coating method. The coated papers were then dried in an oven at 105 ± 2 °C for 15 min. All the coated papers had a constant coated weight of 4 ± 0.5 g/m^2. The morphology of both the surface and the cross-section of the paper samples were examined through a field emission scanning electron microscope (FE-SEM) (Su8020, Hitachi, Tokyo, Japan), at an accelerating voltage of 5 kV. The paper samples were sputtered with a platinum coating of 10 nm thickness.

2.5. Antifungal Activity of the Coated Paper

The antifungal activity of the coated paper, against a fungal stain of *C. gloeosporioides*, was evaluated using the disc diffusion method. Briefly, a PDA disc of diameter 6 mm, containing *C. gloeosporioides*, was placed on the surface of a PDA plate (90 mm diameter) using a sterile cork borer. The coated paper discs, of diameter 6 mm, were placed at the center of the PDA medium. This plate was incubated at 25 °C for 7 days and the growth diameter was measured 3, 5, and 7 days after the incubation, with the experiment being repeated five times. The percentage inhibition was calculated using the equation:

$$\text{Inhibition } (\%) = (A - B/A) \times 100 \tag{1}$$

where *A* is the fungal colony radius of the control plate containing the PDA without the paper sample and *B* is the colony radius in the test plate containing the PDA and paper sample.

2.6. Basis Weight and Thickness

Basis weight and thickness of the uncoated and coated papers was measured according to ISO 536:2012 [19] and ISO 534:2011 standards [20], respectively. The weight of 10 individual papers was measured and the mean values were calculated. The thickness of the papers was measured using a micrometer (Lorentzen & Wettres, Stockholm, Sweden). Each paper was randomly measured at five different places and the mean thickness of a single paper was calculated.

2.7. Tensile and Bursting Test

A universal testing machine (Vantage NX, Thwing-Albert Instrument Co. Ltd., Philadelphia, PA, USA) was used to test the tensile strength, according to the ISO 1924-2:2008 standard [21]. The gauge length was 10 cm and the crosshead speed set at 50 mm/min. The papers were cut to a width of 15 ± 0.1 mm and length of 180 ± 1 mm. For the bursting strength was tested using a burst test machine (MTA-2000, Regmed Indústria Técnica de Precisão Ltda., Osasco, Brazil), according to the ISO 2758:2001 standard [22]. The measurement was done on 10 replicates of sample papers.

2.8. Water Absorptiveness

The water absorptiveness of the uncoated and coated papers was determined using the Cobb method according to the ISO 535:1991 standard [23]. The papers were cut into squares of size 14 × 14 cm^2 and clamped inside the ring of a Cobb tester, having an area of 100 m^2. Into the ring, 100 mL of distilled water was poured the water was absorbed for 120 s. The excess water was then poured out and a wet paper was placed between the blotting papers in order to remove the excess surface water on the paper. The Cobb value was measured in terms of the amount of distilled water in g/m^2 and the experiment was repeated 10 times.

2.9. Antifungal Activity of the Coated Paper

The cut orchid flowers (*Dendrobium sonia*) were collected from Siamtaiyoo farm Co. Ltd. (Samut Sakhon, Thailand) to test the antifungal efficacy of the paper. Healthy flowers were selected based on a long stem (45 cm), flower with approximately 7 ± 1 blooms, and 5 ± 1 buds per stem (export quality grade). Boxes of dimensions 52 cm × 40 cm × 60 cm (height × width × length), generally used for shipping the flowers, size of, were obtained from Siamtaiyoo farm Co. Ltd. (Samut Sakhon, Thailand). AgNPs coated sample papers were placed in the entire inside of the boxes, using a double sided tape. Forty stems of freshly cut orchid flowers were prepared for the packaging test, and placed in boxes with and without AgNPs coating. From each box, one flower bloom was selected and its anther was wounded by puncturing with a sterilized pin. Twenty µL of spore suspension (106 spore/mL) was then dropped into the wound. The sample boxes were stored at a room temperature of 25 ± 2 °C and 50% RH for 7 days. Table 1 shows the treatments used during the packaging test, with the experiment repeated five times. The percentage infection was calculated from the equation:

$$\text{Infection (\%)} = (B/A) \times 100 \tag{2}$$

where A is all the orchid flower blooms and B is the number of orchid flower blooms infected with the fungi.

Table 1. The treatments used during the packaging test.

Treatment	Packaging	Pulsing Solution
T_0	Uncoated	Distilled water
T_1	Uncoated	8-HQS 225 ppm + AgNO$_3$ 30 ppm + Sucrose 4%
T_2	Uncoated	8-HQS 225 ppm + AgNPs 20 ppm + Sucrose 4%
T_3	AgNPs coated	8-HQS 225 ppm + AgNO$_3$ 30 ppm + Sucrose 4%
T_4	AgNPs coated	8-HQS 225 ppm + AgNPs 20 ppm + Sucrose 4%

2.10. Statistical Analysis

All the data were statistically analyzed using a completely randomized design (CRD). A one way analysis of variance (ANOVA) was performed and the means were compared in each treatment, using the Duncan's new multiple ranges test (DMRT) (at a significance level of 0.05).

3. Results

3.1. Synthesis of AgNPs

The synthesis of AgNPs was carried out using the chemical reduction of NaBH$_4$ and Na$_3$C$_6$H$_5$O$_7$. The formation of light yellow colored AgNPs in an aqueous solution was determined by UV-Vis spectroscopy, set to the absorbance mode of the surface plasmon resonance (SPR) peak. It was observed that the SPR peak in the UV-visible spectrum of the AgNPs was at a wavelength of 403 nm (Figure 1a). Generally, the absorption of AgNPs depends on the size of the particles [24]. Previous studies have reported that the absorption spectrum peaks at wavelengths between 400–430 nm and can be attributed to the size of AgNPs, which ranges between 20 and 60 nm [18,25,26]. The particle size of AgNPs was measured and further confirmed by an atomic force microscope (AFM). The size of synthesized AgNPs varied between 20 and 70 nm, with an average of 47 ± 9.06 nm, and were topologically spherical in shape (Figure 1a,b).

3.2. Antifungal Activity of Coated Paper

In this study, the antifungal activity of paper against *C. Gloeosporioides*, coated with 20 and 50 ppm of AgNPs, was investigated by observing the growth of the fungal colony diameter. Results are shown in Figure 2 and Table 2. As seen in the figure, the growth of fungi in a paper coated with 50 ppm AgNPs was significantly inhibited ($p < 0.05$), compared with a paper coated with 20 ppm, after an incubation period of 2 days. According to these results, a 50 ppm AgNP coating strongly reduced the fungal growth.

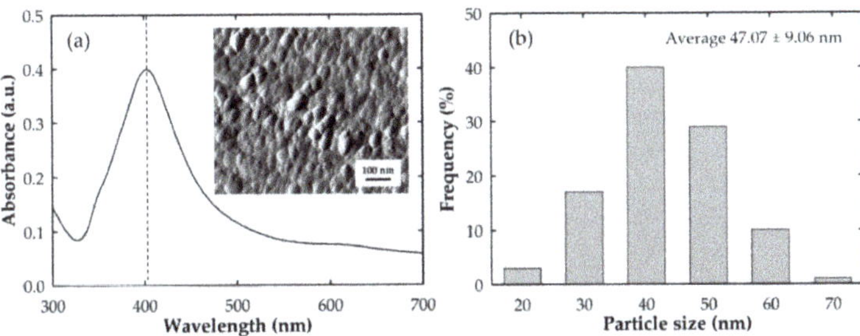

Figure 1. (a) UV–visible spectrum as measured by UV-Vis spectroscopy with an atomic force microscope (AFM) image (1×1 μm^2) of the synthesized AgNPs and (b) particle size distribution of the AgNPs.

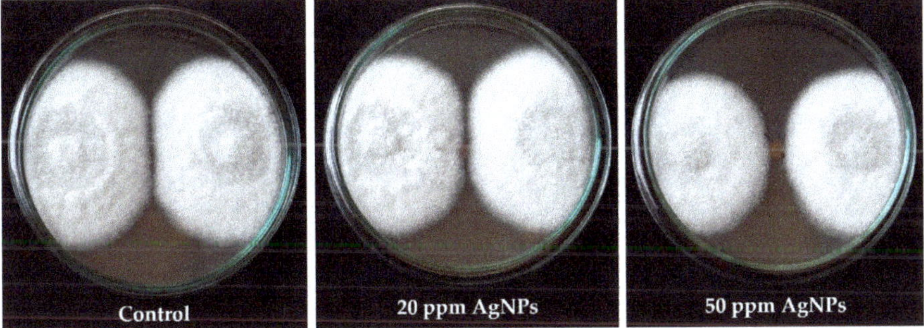

Figure 2. Antifungal activity in a control plate (no AgNP coating) and test plate with different concentrations of AgNPs coated paper, for 7 days of incubation.

Table 2. Inhibition efficacy (% relative to control) of AgNPs coated paper (20 and 50 ppm) to *Colletotrichum Gloeosporioides* fungi.

	Inhibition (%Relative to Control)	
Days	AgNPs Coated Paper (ppm)	
	20	50
1	ND	ND
2	12.00 ± 2.70 [a]	36.00 ± 2.18 [b]
3	3.92 ± 1.79 [a]	18.95 ± 2.73 [b]
4	4.88 ± 2.44 [a]	11.22 ± 2.99 [b]
5	4.26 ± 1.73 [a]	10.08 ± 2.21 [b]
6	2.72 ± 1.42 [a]	6.80 ± 3.27 [b]
7	0.00 ± 0.00 [a]	3.33 ± 1.83 [b]

Data are presented as mean $\pm$ standard deviation, followed by the same letters in the row, indicating that the numbers are not significantly different ($p > 0.05$, based on Duncan's new multiple ranges test (DMRT)), ND–Not determined.

3.3. Morphology of Uncoated and Coated Paper

FE-SEM micrographs of the surface of an uncoated paper indicated that the surface was rough and had a porous fibrous structure, as shown in Figure 3a. After coating the paper with a solution of AgNPs, at about 4.0 g/m^2 of coating weight, the paper showed a relatively smoother surface compared to the uncoated paper (Figure 3b). This smoothness is probably a result of the coating solution filling the pores on the entire surface of the paper, making it homogeneous (Figure 3c) [27,28].

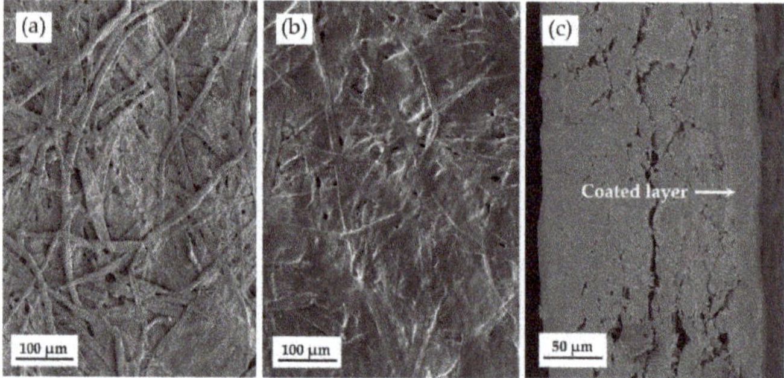

Figure 3. Field emission scanning electron microscope (FE-SEM) micrographs of (**a**) uncoated paper, (**b**) paper coated with AgNPs, and (**c**) cross-section of the surface coated with a layer of AgNPs.

3.4. Basis Weight and Thickness

Table 3 shows the basis weight, thickness, coated weight, and coating thickness of an uncoated and coated paper. The uncoated paper had a basis weight of 134 g/m^2 and a thickness of 174 µm, which were used as a baseline value. When the paper was coated with a coating bar (No. 3), the basis weight and thickness was 137 g/m^2 and 180 µm, respectively. This indicated that the weight of the coated paper increased by 4 g/m^2.

Table 3. The basis weight, thickness, coated weight, and coating thickness of the uncoated and coated paper.

Properties	Paper	
	Uncoated	Coated
Basis weight (g/m^2)	133.88 ± 1.25 [a]	137.61 ± 1.34 [b]
Thickness (µm)	174.28 ± 0.95 [a]	180.00 ± 0.72 [b]
Coating weight (g/m^2)	ND	3.73 ± 0.32
Coating thickness (µm)	ND	5.72 ± 1.15

Data are presented as mean ± standard deviation, followed by the same letters in the row, indicating that the numbers are not significantly different ($p > 0.05$, based on the DMRT); ND–Not determined.

3.5. Tensile and Burst Strength

The tensile and burst strengths of uncoated and coated paper are shown in Table 4. The tensile index of the uncoated paper along the machine direction (MD) and cross machine direction (CD) was 56.33 ± 2.33 and 23.74 ± 1.02 N·m/g, respectively. For the coated paper, the tensile index in MD and CD was 56.02 ± 3.56 and 25.94 ± 0.98 N·m/g, respectively. The tensile index significantly increased with increasing coating weight in CD ($p < 0.05$), while the tensile index did not change in MD. The bursting index of the uncoated paper was 2.79 ± 0.13 kPa·m^2/g, which significantly increased to 3.08 ± 0.08 kPa·m^2/g ($p < 0.05$) when the paper was coated. This was due to the solution creating an excellent film on surface of the paper, which increased the bursting strength [29].

3.6. Water Absorptiveness

The water resistance of an uncoated and coated paper was determined using the Cobb test, and are shown in Figure 4. The resistance of the top side of a coated paper was significantly different ($p < 0.05$) compared to an uncoated paper, and the value decreased from 34.97 ± 1.35 g/m^2 for the uncoated paper to 32.81 ± 0.86 g/m^2 for the coated paper. The water absorptiveness of the bottom side of an uncoated and coated paper was not significantly different ($p > 0.05$), with the value decreasing from 39.82 ± 1.58 g/m^2 for the uncoated paper to 38.65 ± 0.78 g/m^2 for the coated paper. The results indicate that the starch can reduce the water absorptiveness and can effectively improve the moisture barrier properties of hydrophilic films [30,31].

Table 4. The bursting and tensile index of the uncoated and coated paper.

Properties	Paper	
	Uncoated	Coated
Tensile index (Nm/g)	–	–
MD	56.33 ± 2.33 [a]	56.02 ± 3.56 [a]
CD	23.74 ± 1.02 [a]	25.94 ± 0.98 [b]
Bursting index (kPa·m^2/g)	2.79 ± 0.13 [a]	3.08 ± 0.08 [b]

Data are presented as mean $\pm$ standard deviation followed by the same letters in the row, indicating that the numbers are not significantly different ($p > 0.05$, based on the DMRT).

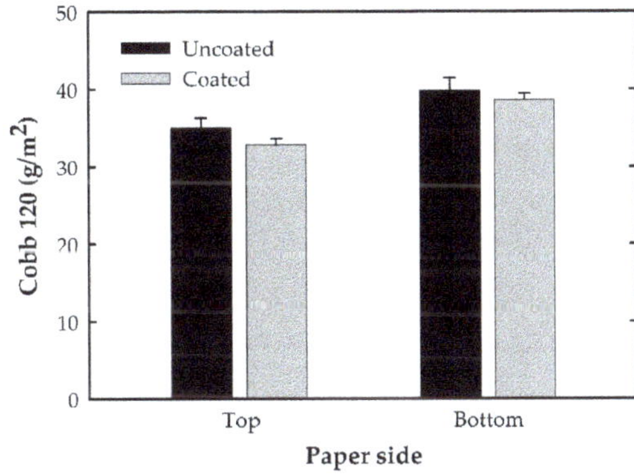

Figure 4. Water absorptiveness values for the uncoated and coated paper as obtained from the Cobb's test.

3.7. Effect of AgNP Coating on C. Gloeosporioides

Table 5 shows the percentage of infected orchid flowers during the packaging test, conducted for 7 days. The highest infection percentage was calculated at 40% for the T_0 treatment. While T_3 and T_4 treatments resulted in a lower infection percentage at 12.5% and significant difference ($p < 0.05$) compared to T_0, T_1, and T_2 treatments. Therefore, a packaging coated with 50 ppm of AgNPs, combined with a pulsing solution (containing AgNO$_3$ and AgNPs), could inhibit the growth of *C. gloeosporioides* mycelium. Photographs of orchid flowers and their respective anthers are shown in Figure 5, used to further determine the hyphae of filamentous of *C. gloeosporioides* on the anther. Hyphae formations were found for T_0, T_1, and T_2 treatments. On the other hand, no *C. gloeosporioides* hyphae formation was observed on the anther for T_3 and T_4 treatments. This validated the speculation that a paper, coated with a concentration of 50 ppm of AgNPs, has good antifungal activity.

Table 5. Percentage of infected cut orchid flowers during the packaging test conducted for 7 days.

Treatments	Packaging	Pulsing Solution	Infection [1] (%)
T_0	Uncoated	Distilled	40.00 ± 0.00 [a]
T_1	Uncoated	8-HQS 225 ppm + AgNO$_3$ 30 ppm + Sucrose 4%	29.17 ± 1.44 [a]
T_2	Uncoated	8-HQS 225 ppm + AgNPs 20 ppm + Sucrose 4%	31.67 ± 3.82 [a]
T_3	AgNPs coated	8-HQS 225 ppm + AgNO$_3$ 30 ppm + Sucrose 4%	12.50 ± 0.00 [b]
T_4	AgNPs coated	8-HQS 225 ppm + AgNPs 20 ppm + Sucrose 4%	12.50 ± 0.00 [b]

[1] Average $\pm$ standard deviation followed by the same letters in column indicating that the numbers are not significantly different ($p > 0.05$, based on DMRT).

Figure 5. Photographs of cut orchid flowers (left) and respective anthers (right), after being stored for 7 days during the packaging test. Hyphae formations can be observed for T_0, T_1, and T_2 treatments. On the other hand, no *C. gloeosporioides* hyphae formation is observed on the anther for T_3 and T_4 treatments.

4. Conclusions

The AgNPs, synthesized in this study, which were spherical in shape and 47 nm in diameter, were prepared using chemical reduction and a stabilizing solution at 90 °C. Results from the UV-visible spectrography and AFM analysis confirmed the presence of AgNPs and their topology. The paper coated with 50 ppm of AgNPs had an excellent antifungal activity against *C. gloeosporioides*, under culture conditions. The AgNPs-starch coating increased the water resistance, tensile strength in CD, and bursting strength, of the paper. The paper coated with AgNPs, was successful in decreasing the infection caused by *C. gloeosporioides*, on cut orchid flowers. Therefore, AgNP coated paper has a great potential to be used widely in the packaging industry for producing boxes which can inhibit the *C. gloeosporioides* infection that plagues cut orchid flowers and other crops.

Author Contributions: Conceptualization, B.N. and B.P.; Methodology, B.N.; Software, B.N. and B.P.; Validation, B.N., S.P., S.K., and B.P.; Formal Analysis, B.N. and B.P.; Resources, B.N. and B.P.; Data Curation, S.P. and S.K.;

Writing-Original Draft Preparation, B.N. and B.P.; Writing-Review & Editing, B.N. and B.P.; Visualization, B.N. and B.P.; Supervision, B.P.; Project Administration, B.N.; Funding Acquisition, B.N. and B.P.

Funding: This work was supported by the Graduate Scholarship from the National Research Council of Thailand (NRCT2561001).

Acknowledgments: We are grateful to Jantana Praiboon for her helpful technical assistance.

Conflicts of Interest: The authors declare no conflict of interest.

References

1. Chowdappa, P.; Chethana, C.S.; Bharghavi, R.; Sandhya, H.; Pant, R.P. Morphological and molecular characterization of *Colletotrichum gloeosporioides* (Penz) Sac. isolates causing anthracnose of orchids in India. *Biotechnol. Bioinf. Bioeng.* **2012**, *2*, 567–572.

2. Office of Agricultural Economic. Exporting Value of Orchid Cut Flower in Thailand on 2015–2017. Available online: http://oldweb.oae.go.th/production.html (accessed on 11 January 2019).

3. Thammasiri, K. Thai orchid genetic resources and their improvement. *Horticulturae* **2016**, *2*, 9. [CrossRef]

4. Wittayawannakul, W.; Tassakorn, T.; Bhasabutra, T. Postharvest control of black anther disease in dendrobium. *Agric. Sci. J.* **2012**, *43*, 584–587.

5. Tassakorn, T.; Bhasabutra, T.; Wittayawannakul, W. Chemical Control of Black Anther Disease on Dendrobium Orchids. Available online: http://www.doa.go.th/research/showthread.php?tid=1781 (accessed on 11 March 2018).

6. Chowdappa, P.; Gowda, S.; Chethana, C.S.; Madhura, S. Antifungal activity of chitosan-silver nanoparticle composite against *Colletotrichum gloeosporioides* associated with mango anthracnose. *Afr. J. Microbiol. Res.* **2014**, *8*, 1803–1812.

7. Junqueira-Gonçalves, M.P.; Alarcón, E.; Niranjan, K. Development of antifungal packaging for berries extruded from recycled PET. *Food Control* **2013**, *33*, 455–460. [CrossRef]

8. Rudra, S.G.; Singh, V.; Jyoti, S.D.; Shivhare, U.S. Mechanical properties and antimicrobial efficacy of active wrapping paper for primary packaging of fruits. *Food Biosci.* **2013**, *3*, 49–58. [CrossRef]

9. Lamsal, K.; Kim, S.W.; Jung, J.H.; Kim, Y.S.; Kim, K.S.; Lee, Y.S. Application of silver nanoparticles for the control of *Colletotrichum* species in vitro and pepper anthracnose disease in field. *Mycobiology* **2011**, *39*, 194–199. [CrossRef] [PubMed]

10. Aguilar-Méndez, M.A.; Martín-Martínez, E.S.; Ortega-Arroyo, L.; Cobián-Portillo, G.; Sánchez-Espíndola, E. Synthesis and characterization of silver nanoparticles: Effect on phytopathogen *Colletotrichum gloesporioides*. *J. Nanoparticle Res.* **2013**, *13*, 2525–2532. [CrossRef]

11. Nokkrut, B.; Pisutpiched, S.; Puangsin, B. Effect of quaternary ammonium compounds and zinc pyrithione on antimicrobial properties in packaging containers for orchid cut flowers. *Thai J. For.* **2017**, *36*, 136–144.

12. Chaichompoo, W.; Phichai, K. Effect of plant extract on growth inhibition of *Colletotrichum* sp. *Res. J.* **2010**, *3*, 18–25.

13. Kang, H.C.; Park, Y.H.; Go, S.J. Growth inhibition of a phytopathogenic fungus, *Colletotrichum* species by acetic acid. *Microbiol. Res.* **2003**, *158*, 321–326. [CrossRef] [PubMed]

14. Shi, X.; Li, B.; Qin, G.; Tian, S. Mechanism of antifungal action of borate against *Colletotrichum gloeosporioides* related to mitochondrial degradation in spores. *Postharvest Biol. Technol.* **2012**, *67*, 138–143. [CrossRef]

15. Bussaman, P.; Namsena, P.; Rattanasena, P.; Chandrapatya, A. Effect of crude leaf extracts on *Colletotrichum gloeosporioides* (Penz.) Sacc. *Psyche A J. Entomol* **2012**, *1*, 309046.

16. Echegoyen, Y.; Nerín, C. Nanoparticle release from nano-silver antimicrobial food containers. *Food Chem. Toxicol.* **2013**, *62*, 16–22. [CrossRef]

17. Carbone, M.; Donia, D.T.; Sabbatella, G.; Antiochia, R. Silver nanoparticles in polymeric matrices for fresh food packaging. *J. King Saud Univ. Sci.* **2016**, *28*, 273–279. [CrossRef]

18. Agnihotri, S.; Mukherji, S.; Mukherji, S. Size-controlled silver nanoparticles synthesized over the range 5–100 nm using the same protocol and their antibacterial efficacy. *RSC Adv.* **2014**, *4*, 3974–3983. [CrossRef]

19. *ISO 536 Paper and Board-Determination of Grammage*; International Organization for Standardization: Geneva, Switzerland, 1995.

20. *ISO 534 Paper and Board-Determination of Thickness, Density and Specific Volume*; International Organization for Standardization: Geneva, Switzerland, 2005.

21. *ISO 1924-2 Paper and Board-Determination of Tensile Properties Part 2 Constant Rate of Elongation Method*; International Organization for Standardization: Geneva, Switzerland, 2008.

22. *ISO 2758 Paper Determination of Bursting Strength*; International Organization for Standardization: Geneva, Switzerland, 2001.

23. *ISO 535 Paper and Board-Determination of Water Absorptiveness-Cobb Method*; International Organization for Standardization: Geneva, Switzerland, 1991.

24. Pacioni, N.L.; Borsarelli, C.D.; Rey, V.; Veglia, A.V. Synthetic routes for the preparation of silver nanoparticles. In *Silver Nanoparticle Applications*; Alarcon, E.I., Griffith, M., Udekwu, K.I., Eds.; Springer International Publishing: Basel, Switzerland, 2015; Volume 1, pp. 13–45.

25. Paramelle, D.; Sadovoy, A.; Gorelik, S.; Free, P.; Hobley, J.; Fernig, D.G. Rapid method to estimate the concentration of citrate capped silver nanoparticles from UV-visible light spectra. *Analyst* **2014**, *139*, 4855–4861. [CrossRef]

26. Tomaszewska, E.; Soliwoda, K.; Kadziola, K.; Tkacz-Szczesna, B.; Celichowski, G.; Cichomski, M.; Szmaja, W.; Grobelny, J. Detection limits of DLS and UV-Vis spectroscopy in characterization of polydisperse nanoparticles colloids. *J. Nanomater.* **2013**, 60. [CrossRef]

27. Li, Y.; Zhang, J.; Li, H.; Gu, W.J.; He, B. Analysis of the coating surface properties of coated paper. In Proceedings of the 2015 International Conference on Materials, Environmental and Biological Engineering (MEBE 2015), Guilin, China, 28–30 March 2015; pp. 250–253.

28. Amini, E.; Azadfallah, M.; Layeghi, M.; Talaei-Hassanloui, R. Silver-nanoparticle-impregnated cellulose nanofiber coating for packaging paper. *Cellulose* **2016**, *23*, 557–570. [CrossRef]

29. Bruun, S.E. Starch. In *Pigment Coating and Surface Sizing of Surface*; Lehtnen, E., Suomen Paperi-insinöörien Yhdistys, Technical Association of the Pulp and Paper Industry, Eds.; Fapet Oy: Jyväskylä, Finland, 2000; Volume 1, pp. 241–249.

30. Butinaree, S.; Jinkarn, T.; Yoksan, R. Effects of biodegradable coating on barrier properties of paperboard food packaging. *J. Met. Mater. Miner.* **2008**, *18*, 219–222.

31. Biricik, Y.; Sonmez, S.; Ozden, O. Effects of surface sizing with starch on physical strength properties of paper. *Asian J. Chem.* **2011**, *23*, 3151–3154.

 *coatings*

Article

Polymeric Antimicrobial Coatings Based on Quaternary Ammonium Compounds

Denisa Druvari [1], Nikos D. Koromilas [1,2], Vlasoula Bekiari [3], Georgios Bokias [1,2] and Joannis K. Kallitsis [1,2,*]

[1] Department of Chemistry, University of Patras, GR–26504 Patras, Greece; druvari@upatras.gr (D.D.); nikoskoromil@upatras.gr (N.D.K.); bokias@upatras.gr (G.B.)

[2] FORTH/ICE-HT, Stadiou Str., P.O. Box 1414, GR–26504 Rio-Patras, Greece

[3] Department of Fisheries and Aquaculture Technology, Technological Educational Institute of Western Greece, 30200 Mesologhi, Greece; mpekiari@teimes.gr

[*] Correspondence: j. kallitsis@upatras.gr; Tel.: +30-261-096-2952; Fax: +30-261-099-7122

Academic Editor: Sami Rtimi

Received: 23 November 2017; Accepted: 20 December 2017; Published: 23 December 2017

Abstract: Biocidal coatings that are based on quaternized ammonium copolymers were developed after blending and crosslinking and studied as a function of the ratio of reactive groups and the type of biocidal groups, after curing at room temperature or 120 °C. For this purpose, two series of copolymers with complementary reactive groups, poly(4-vinylbenzyl chloride-co-acrylic acid), P(VBC co AAx), and poly(sodium 4-styrenesulfonate-co-glycidyl methacrylate), P(SSNa-co-GMAx), were synthesized via free radical copolymerization and further modified resulting in covalently bound (4-vinylbenzyl dimethylhexadecylammonium chloride, VBCHAM) and electrostatically attached (hexadecyltrimethylammonium 4-styrene sulfonate, SSAmC$_{16}$) units. The crosslinking reaction between the carboxylic group of acrylic acid (AA) and the epoxide group of glycidyl methacrylate (GMA) of these copolymers led to the stabilization of the coatings through reactive blending. The so developed coatings were cured at room temperature and 120 °C, and then immersed in ultra-pure water and aqueous NaCl solutions at various concentrations for a time period up to three months. Visual inspection of the integrity of the materials coated onto glass slides, gravimetry, scanning electron microscopy (SEM) characterization, as well as the determination of total organic carbon (TOC) and total nitrogen (TN) of the solutions, were used to investigate the parameters affecting the release of the materials from the coatings based on these systems. The results revealed that curing temperature, complementary reactive groups' content, and type of antimicrobial species control the release levels and the nature of releasable species of these environmentally-friendly antimicrobial coatings.

Keywords: antimicrobial action; quaternary ammonium groups; acrylic acid; glycidyl methacrylate; crosslinking reaction; coating

1. Introduction

Research on antimicrobial polymeric materials and surfaces is intensive [1–7], since such materials are important for diverse applications, like health care/biomedical devices, food packaging, agriculture/aquaculture, marine biofouling, etc. [8–11]. Among the numerous examples of antimicrobial polymers that were evaluated for such potential applications [12–14], polymers based on quaternary ammonium groups [15–18] are possibly the most widely studied, concerning synthetic biocidal polymeric materials [19].

A critical issue regarding infection-related health is the formation of bacterial biofilms on surfaces. To overcome this problem, an urgent need for the development of coatings that are able to prevent biofilm formation or eliminate already-constructed biofilms is well established [20–22]. Efficient antimicrobial coatings can prevent bacteria adhesion or kill the bacteria before or after contact with the surface, although many strategies include both mechanisms [23–26].

To better tune biocidal activity and duration, research nowadays focuses on dual-action biocidal materials and surfaces, i.e., materials with a simultaneous contact-killing and released-based biocidal action. The contact-killing action is usually based on quaternary ammonium groups that are covalently incorporated into a polymeric chain or polymeric substrate [27–29], whereas the released-based biocidal action is often introduced through the use of biocidal agents capable of leaching from the polymeric surfaces or materials, such as metallic/inorganic nanoparticles, antibiotics, or others [30].

Having this in mind, our research group is focusing on the potential use of quaternary ammonium groups with contact-based action, released-based action, as well as the combination of both actions. Thus, hexadecyltrimethylammonium (cetyl trimethylammonium) cations (AmC$_{16}$) electrostatically bound onto an anionic poly(styrene sulfonate) (PSS) backbone have been evaluated as potential released-based polymeric biocidal materials, while polymers of 4-vinylbenzyl chloride (VBC) modified with *N,N*-dimethylhexadecylamine (HAM) have been developed to achieve the contact-based action [31]. Moreover, random or block copolymers of the respective units, hexadecyltrimethylammonium styrene sulfonate (SSAmC$_{16}$) and 4-vinylbenzyl dimethylhexadecylammonium chloride (VBCHAM), have been designed for the combination of both actions [32].

Recently, we have shown that we can take advantage of the reactive blending concept, in order to prepare self-standing crosslinked membranes containing both released-based hexadecyltrimethylammonium groups and contact-based VBCHAM biocidal units [33]. Thus, polymeric precursors of the two biocidal species with complementary chemical functions, e.g., carboxylic groups of acrylic acid (AA) units and epoxide groups of glycidyl methacrylate (GMA) units, were initially synthesized. Blends of these copolymers, P(SSAmC$_{16}$-co-GMAx) and P(VBCHAM-co-AAx) were then formed, and were cured at solid state at the desired temperature, leading to the final crosslinked membranes (Scheme 1), as a consequence of the reaction between the carboxylic and the epoxide groups. These membranes presented strong antimicrobial activity against *S. aureus* and *P. Aeruginosa*, while when applied as coatings on aquaculture nets exhibited high antifouling action as compared to blank net [33]. An interesting observation in that work was that the observed release in salt solution was maintained in much lower levels than the release in pure water, offering an additional advantage for potential antifouling applications in sea water.

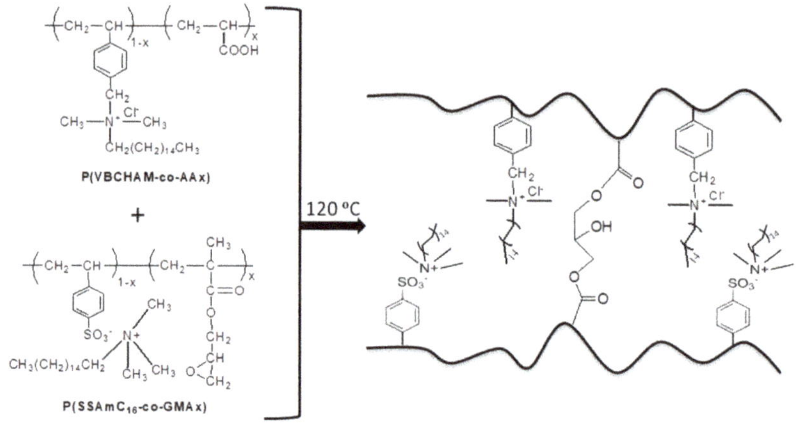

Scheme 1. Reaction between P(VBCHAM-co-AAx) and P(SSAmC$_{16}$-co-GMAx) copolymers after curing at 120 °C.

Motivated by the aforementioned encouraging initial findings, our aim in the present work is to investigate in more detail this methodology by using different copolymers' composition and blending ratio, in order to get a deeper understanding on the structural factors affecting the release behavior. Such knowledge is a prerequisite in order to optimize the efficacy and duration of the antifouling action of these novel polymeric biocidal coatings.

2. Materials and Methods

2.1. Materials

The monomers glycidyl methacrylate (GMA), sodium 4-styrene sulfonate (SSNa), acrylic acid (AA) and 4-vinylbenzyl chloride (VBC), the initiator azobisisobutyronitrile (AIBN), the surfactant hexadecyltrimethylammonium (cetyl trimethylammonium) bromide (AmC$_{16}$), the amine *N,N*-dimethylhexadecylamine (HAM), the salt NaCl; as well as deuterium oxide (D$_2$O), and deuterated chloroform (CDCl$_3$) were purchased from Aldrich, and were used as received. The solvents *N,N*-dimethylformamide (DMF), dimethylsulfoxide (DMSO), chloroform (CHCl$_3$), hexane and acetone were purchased from Fischer and used as received. Micro glass slides with size ± 26 mm × 76 mm were acquired from Sigma-Aldrich. Ultra-pure water was obtained by means of an SG apparatus water purification unit.

2.2. Synthesis of the Precursors

The copolymers poly(4-vinylbenzyl chloride-co-acrylic acid) and poly(sodium 4-styrenesulfonate-co-glycidyl methacrylate) were synthesized through free radical polymerization in CHCl$_3$ at 70 °C and DMF/pure H$_2$O (50/50) or DMSO at 80 °C, respectively, using AIBN as initiator. These copolymers (Table 1) will be denoted as P(VBC-co-AAx) and P(SSNa-co-GMAx), where x is the mole fraction of AA and GMA units in the copolymer, as determined by the ^{1}H NMR characterization in CDCl$_3$ and D$_2$O, respectively. The experimental procedures for the synthesis and characterization of copolymers have been described previously [31,33].

Table 1. Characterization results for the P(VBCHAM-co-AAx) and P(SSAmC$_{16}$-co-GMAx) complementary antimicrobial copolymers.

Precursors	Feed Composition % (mol AA or GMA)	^{1}H NMR Composition % (mol AA or GMA)	M_w	PDI	Antimicrobial Copolymers
P(VBC-co-AA7)	10	7	22,000	1.7	P(VBCHAM-co-AA7)
P(VBC-co-AA20)	20	20	28,200	2.7	P(VBCHAM-co-AA20)
P(SSNa-co-GMA6)	5	6	27,800	1.9	P(SSAmC$_{16}$-co-GMA6)
P(SSNa-co-GMA20)	15	20	12,200	1.8	P(SSAmC$_{16}$-co-GMA20)

2.3. Introduction of Antimicrobial Species

2.3.1. Covalently Bound Antimicrobial Groups

The quaternization process of P(VBC-co-AAx) copolymers has been described previously [31]. Briefly, the copolymers were dissolved in CHCl$_3$ and quaternized with an excess of *N,N*-dimethylhexadecylamine (HAM) at 60 °C for 48 h. The quaternized polymers, denoted P(VBCHAM-co-AAx), were recovered by precipitation in acetone, thoroughly washed with hexane, and dried in a vacuum oven at 60 °C for 24 h.

2.3.2. Electrostatically Attached Antimicrobial Groups

The introduction of electrostatically attached quaternary ammonium cations on the P(SSNa-co-GMAx) copolymers or PSSNa homopolymer was achieved through an ion exchange reaction in aqueous solution between the sodium cations of SSNa units and an excess of quaternary hexadecyltrimethylammonium cations (AmC_{16}). The final products, denoted P(SSAmC$_{16}$-co-GMAx), were obtained through filtration, washed thoroughly with ultra-pure water and dried in a vacuum oven at 60 °C for 24 h. The experimental procedures are described in detail elsewhere [33,34].

2.4. Preparation of Antimicrobial Coatings

Mother solutions of series of the two copolymers P(VBCHAM-co-AAx) and P(SSAmC$_{16}$-co-GMAx) were prepared in CHCl$_3$ at a 10% (w/v) concentration and left overnight at room temperature under mild stirring. Afterwards, the mother solutions of the complementary copolymers were mixed at various compositions. The compositions of P(VBCHAM-co-AAx) and P(SSAmC$_{16}$-co-GMAx) were expressed as weight/weight ratio (w/w) and set at the desired values (20/80, 40/60, 85/15) (Table 2). Subsequently, an appropriate volume of each mixture (1 mL) was incorporated onto the surface of micro glass slides and was left at room temperature for 24 h until complete solvent evaporation. Then, the coated slides were cured at 120 °C. For comparison reasons, uncured coated slides were used.

Table 2. Composition and curing conditions of the P(VBCHAM-co-AAx) and P(SSAmC$_{16}$-co-GMAx) complementary antimicrobial copolymers coated onto glass slides.

Complementary Copolymers		Composition, % *w/w*	Curing Temperature and Time of Curing	Polymeric Coating
P(SSAmC$_{16}$-co-GMA20)	P(VBCHAM-co-AA7)	40/60	RT (1 day) 120 °C (1 day)	C1-RT C1-120
P(SSAmC$_{16}$-co-GMA6)	P(VBCHAM-co-AA20)	40/60	RT (1 day) 120 °C (1 day)	C2-RT C2-120
P(SSAmC$_{16}$-co-GMA20)	P(VBCHAM-co-AA20)	40/60 85/15 20/80	RT (1 day) 120 °C (1 day) 120 °C (1 day) 120 °C (1 day)	C3-RT C3-120 C4-120 C5-120

Note: RT, Room temperature.

2.5. Immersion of Coatings in Ultra-Pure Water and Aqueous NaCl Solutions

The uncured and cured coated glass slides were immersed in ultra-pure water and aqueous NaCl solutions at 1 M, 0.5 M, and 0.25 M concentrations for significant time periods up to three months (1 day, 7 days, 30 days, 60 days). The volume was set at 130 mL, until complete coverage of the coated glass slides' surface was achieved. Finally, the immersed coated glass slides were taken out at the specific time period, immersed in ultra-pure water for the removal of NaCl when necessary and dried at room temperature for two days.

2.6. Characterization Techniques

2.6.1. Scanning Electron Microscopy (SEM) Examination

Scanning electron microscopy (SEM, Zeiss SUPRA 35VP instrument equipped with an Energy-dispersive X-ray spectroscopy, EDS, detector, Carl Zeiss AG, Oberkochen, Germany) was performed in order to investigate the polymeric coatings' surface morphologies.

2.6.2. Release Studies

The coated glass slides were immersed in ultra-pure water or aqueous NaCl solutions and left for different time intervals up to three months. The glass slides were removed at the specific time period, washed (in the case of NaCl solutions) and dried. The soluble fraction % (w/w) of the membranes in ultra-pure water or aqueous NaCl solutions was evaluated gravimetrically from the equation soluble fraction % (w/w) = $\frac{|w-w_0|}{w_0}$ %, where W_0 and W are the measured weights of the coatings before and after the immersion, respectively.

2.6.3. Total Organic Carbon (TOC) and Total Nitrogen (TN) Measurements

Simultaneous analyses of TOC and TN were carried out using a Shimadzu TOC analyzer (TOC-VCSH) coupled to a chemiluminescence detector (TNM-1 TN unit). TOC analysis was performed using the Combustion-Infrared method. The principle of this method is that a microportion of the sample is injected into a heated reaction chamber packed with an oxidative catalyst, Pt/Al_2O_3. The organic and inorganic carbon is oxidized to CO_2, and is measured by means of a nondispersive infrared analyzer (NDIR analyzer). TN analysis was performed using the Pyrolysis-Chemiluminescence detection method. Oxidative pyrolysis converts chemically bound nitrogen to nitric oxide (NO). Nitric oxide is contacted with ozone (O_3) to produce metastable nitrogen dioxide (NO_2^*). As NO_2^* decays, the emitted light is detected by a photomultiplier tube.

3. Results and Discussion

The selection of the copolymers that will be used as the blends components was made in such a way that the initial biocidal activity was assured. More specifically, the copolymers with high content of electrostatically attached ammonium groups were used (see Table 1) that were previously tested and showed high biocidal activity in the range of 5–6 logarithmic reduction for all microbial that were tested [31]. The respective copolymers with covalently bound ammonium groups P(VBCHAM-co-AAx) have shown high activity against *E. faecalis* and *P. aureginosa* [31]. Additionally, all of the random copolymers [31] and coatings [33] having both types of ammonium groups in a comparable ratio have shown extremely high biocidal activity, as shown in the previous publications [31,33].

The main goal of the present work is a deeper understanding of the factors affecting the release rate of crosslinked antimicrobial polymeric coatings with dual contact-based and release-based antimicrobial activity. For this purpose, as was shown previously [33], the complementary reactive antimicrobial polymers P(VBCHAM-co-AAx) and P(SSAmC$_{16}$-co-GMAx) were used for the coating development and stabilization. More specifically, new series of copolymers with AA and GMA groups contents from 6% up to 20% (see Table 1) were synthesized. The respective solutions in CHCl$_3$ of blends of the above copolymers were stirred for adequate time until complete homogenization. Then, they were casted on glass slides and the coated glass slides, after treatment either at room temperature (uncured) or at 120 °C (cured), were tested in respect to their release behavior by immersion in aqueous NaCl solutions or ultra-pure water for a period of time up to three months.

In order to understand the behavior, the initial solubility of the copolymers with the highest amount of the reactive groups e.g., P(SSAmC$_{16}$-co-GMA20) and P(VBCHAM-co-AA20), was tested applying the same coating and release procedure in aqueous NaCl 1 M solution and ultra-pure water for 30 days. The results are shown in Figure 1 in terms of soluble fraction % (*w/w*) and total organic carbon (TOC) evolution. As expected, despite the presence of the hydrophilic AA groups, the solubility of hydrophobic P(VBCHAM-co-AA20) is marginal in water and salt solution, as a consequence of the hydrophobic character of VBCHAM units. On the other hand, the copolymer P(SSAmC$_{16}$-co-GMA20) is readily soluble in salt solution, since it is well-known that the addition of electrolytes weakens the interactions between polyelectrolytes and oppositely charged surfactants [35]. Interestingly, at the low polymer concentration that was applied for these studies, this copolymer is also soluble in pure water. This behavior has been also observed for the complexation of CTAB with copolymers of SSNa with methyl methacrylate (MMA) [36] and possibly originates from the copolymer structure of the polyelectrolyte, leading to the disruption of the synergistic character of the polyelectrolyte/surfactant complexation. As an additional step in this initial study, the release of a PSSAmC$_{16}$/P(VBCHAM-co-AA20) 40/60 *w/w* was also followed under similar conditions. The homopolymer PSSAmC$_{16}$ was used in this case, instead of the copolymer, in order to avoid any crosslinking possibility even at room temperature. As seen, a significant fraction of the blend, comparable to the PSSAmC$_{16}$ content of the blend, is solubilised in pure water. Moreover, the soluble fraction in salt solution is very low and comparable to that of P(VBCHAM-co-AA20), besides the high solubility of PSSAmC$_{16}$ in aqueous NaCl 1 M solution. The investigation of these intriguing solubility characteristics, already observed earlier with similar crosslinked membranes [33], is one of the main goals of the present work.

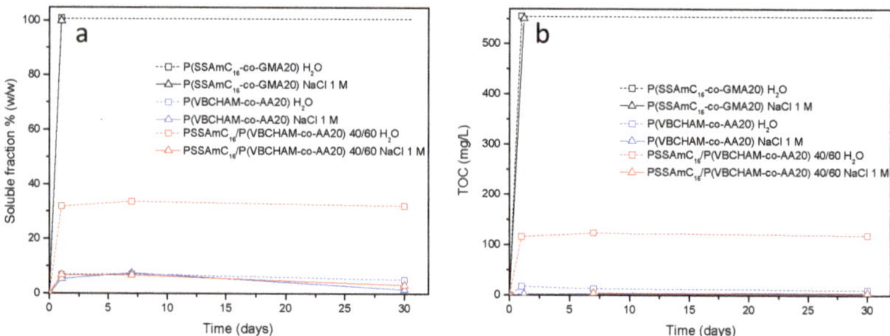

Figure 1. (**a**) Evolution of the soluble fraction % (*w*/*w*) after immersion of P(SSAmC$_{16}$-co-GMA20), P(VBCHAM-co-AA20) and PSSAmC$_{16}$/P(VBCHAM-co-AA20) 40/60 *w*/*w* polymeric coatings in ultra-pure water and aqueous NaCl 1 M solution for different time periods; and, (**b**) Evolution of the total organic carbon (TOC) values of the solutions for the same studies.

Three systems of the complementary antimicrobial polymers were initially investigated (C1-C3). In these systems, the ratio of the complementary reactive units varied significantly, whereas the content of the two types of biocidal units, i.e., releasable SSAmC$_{16}$ groups or immobilized quaternary ammonium VBCHAM units, was rather constant, since they were prepared at a fixed mixing weight ratio (40/60, see Table 2). The polymeric coatings presented different mechanical integrity after their immersion, depending on the curing temperature and the salinity of the aqueous solution. A characteristic example is depicted in Figure 2 for the C3 polymeric coatings, i.e., the uncured (C3-RT) and cured (C3-120) P(SSAmC$_{16}$-co-GMA20)/P(VBCHAM-co-AA20) 40/60 *w*/*w* polymeric coatings. It is evident that in ultra-pure water the C3-RT disintegrate after a few minutes (Figure 2a), whereas the C3-120 detach while preserving its integrity (Figure 2b). This behavior is mostly attributed to the crosslinking effect at 120 °C. On the other hand, both C3-RT and C3-120 polymeric coatings remain intact in the aqueous NaCl 1 M solution (Figure 2c,d).

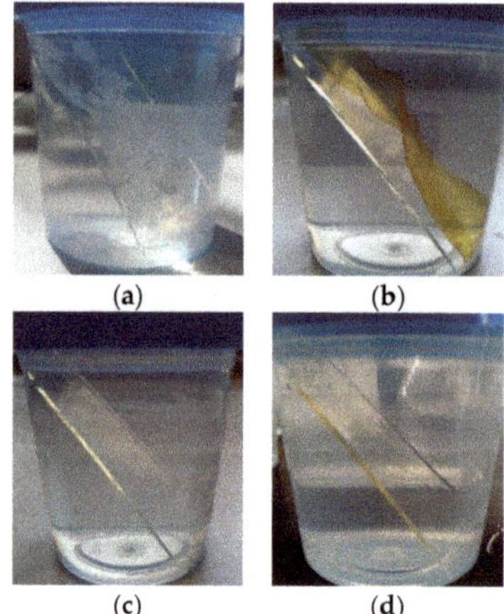

Figure 2. Photos of (**a**,**c**) Uncured C3-RT or (**b**,**d**) Cured C3-120 polymeric coatings, after immersion in ultra-pure water (**a**,**b**) or aqueous NaCl 1 M solution (**c**,**d**) for 7 days.

The release of leachable species of the C1-C3 polymeric coatings in pure water or aqueous NaCl 1 M solution was also followed as a function of time. The evolution of the soluble fraction, determined gravimetrically from the weight change of the coatings, is presented in Figure 3. The observed release trends were also verified from the TOC and TN values determined in the solutions for the same periods (Figures S1 and S2). The dashed lines correspond to the P(SSAmC$_{16}$-co-GMAx) content and represent the upper release limit when considering full solubility for these copolymers and marginal solubility for the complementary P(VBCHAM-co-AAx) copolymers. As seen, both for cured and uncured coatings, the release in water is higher than in NaCl 1 M solution, while the soluble fraction levels do not differentiate substantially with the change of treatment temperature. In addition, the soluble fraction levels are not significantly affected by the chemical composition of the complementary copolymers, i.e., the content of reactive units.

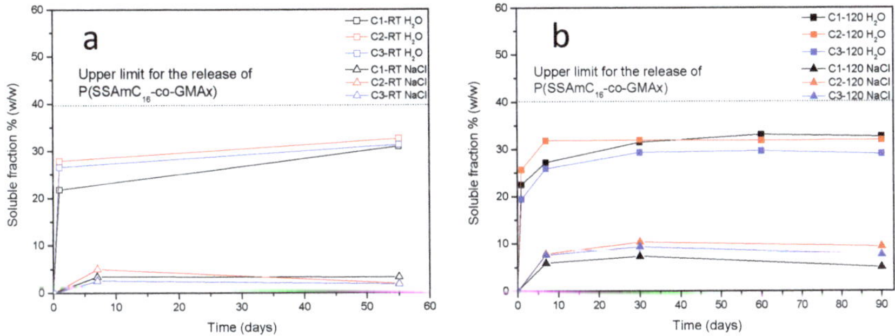

Figure 3. Evolution of the soluble fraction % (*w/w*) after immersion in ultra-pure water and aqueous NaCl 1 M solution for different time periods. (**a**) Uncured C1-RT, C2-RT, C3-RT polymeric coatings, (**b**) Cured C1-120, C2-120, C3-120 polymeric coatings.

These observations, in combination with the behavior presented in Figure 1 for the PSSAmC$_{16}$/P(VBCHAM-co-AA20) coating, indicate that the release of leachable species is not controlled by the possible crosslinking reaction between the complementary reactive units, but it is the result of non-covalent interactions between the two complementary copolymers. However, the crosslinking reaction is a crucial factor, concerning the mechanical integrity of the coatings: the visual inspection of coatings shows a clear difference between the cured and uncured coatings, with the coatings being developed at 120 °C to be more robust. For that reason, the release study for this type of coatings that was performed for three months showed a slow release after the first 10 days, but the quality of the coatings remained unchanged for the whole period of testing.

In order to understand these behaviors in terms of the releseable species [37], the C/N molar ratio was calculated from the TOC/TN ratio and plotted versus time, as shown in Figure 4. The C/N molar ratio is releated either to the cetyl trimethylammonium ions (AmC$_{16}$) or to the whole polymeric chain. As a consequence, the expected values are 19 for the cetyl units (dotted line) and about 27–28 for the whole polymeric chains of both complemetary polymers (dashed area). As seen in this Figure, the release comes in most cases from the liberation of the cetyl trimethylammonium ions (C/N molar ratios about 16–20). The release of these cations in the case of coatings cured at 120 °C is expected in aqueous NaCl 1 M solutions, since they can be ion exchanged with the Na$^+$ cations of the salt, while the polymeric chains cannot be released as they are crosslinked. However, the explanation for the release of cetyl trimethylammonium cations is not straightforward when the same coatings are treated with ultra-pure water. Possibly, in these systems, an ion rearrangement takes place, allowing for the internal complexation of the positive VBCHAM and negative styrene sulfonate ions immobilized onto the different polymeric backbones of the coating. This helps the release of cetyl trimethylammonium cations (from SSAmC$_{16}$ units) with Cl$^-$ anions (from VBCHAM units) as counterions. In the case of the coatings cured at room temperature, the crosslinking conditions are very mild, and crosslinking, if any, is expected to be very low. In fact, in these systems the main trends of the C/N molar ratios are quite similar to those observed for the PSSAmC$_{16}$/P(VBCHAM-co-AA20) 40/60 *w/w* coating (Figure S3). Thus, in the case of this coating and the coatings cured at room temperature, the ion rearrangement described previously could also take place in ultra-pure water, leading to the release of AmC$_{16}$ cations, as evidenced by the observed values of C/N molar ratios (~18–20). In contrast, when these coatings are treated with aqueous NaCl 1 M solutions, high C/N molar ratios are observed (~25–27). In these cases, the release (though low) is made from the whole polymeric chain (PSSAmC$_{16}$ in Figure S3 and P(SSAmC$_{16}$-co-GMAx) in Figure 4). Apparently, when crosslinking cannot be applied (PSSAmC$_{16}$) or the crosslinking conditions are very mild (coatings treated at room temperature), the respective polymers (PSSAmC$_{16}$ or P(SSAmC$_{16}$-co-GMAx)) turn to water-soluble in salt solution, as discussed in Figure 1. In fact, the solubilization of the whole polymeric chain seems to be gradual for the P(SSAmC$_{16}$-co-GMAx)/P(VBCHAM-co-AAx) coatings, since the high C/N molar ratios are observed at large releasing times, whereas these ratios are of the order of 19 for shorter releasing times. Alhough more detailed studies are needed, this observation possibly indicates the potentiality to develop self-eroding biocidal coatings using such materials.

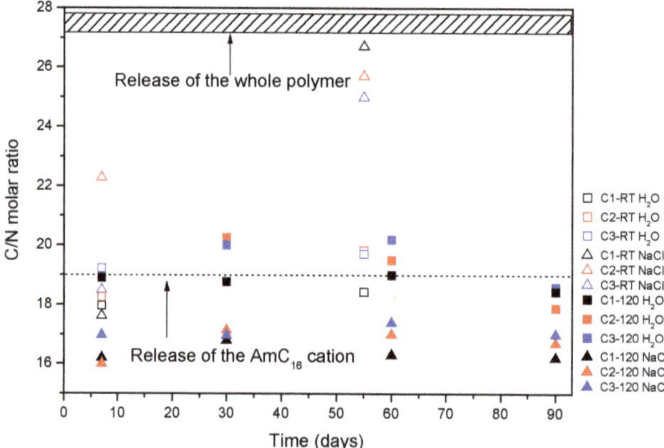

Figure 4. Evolution of the C/N molar ratio, determined from the TOC/ total nitrogen (TN) studies, after immersion in ultra-pure water and NaCl 1 M for different time periods of uncured C1-RT, C2-RT, C3-RT polymeric coatings and cured C1-120, C2-120, C3-120 polymeric coatings.

The influence of the blend composition and more specifically the ratio between the electrostatically and covalently bound quaternary ammonium biocidal species of the initial copolymers on the release properties was examined using the system P(SSAmC$_{16}$-co-GMA20)/P(VBCHAM-co-AA20), after curing at 120 °C. The release characteristics of three coatings with weight compositions 85/15 (C4-120), 40/60 (C3-120), and 20/80 (C5-120) were evaluated for 30 days (Figure 5). The system C5-120 with the lower content of the exchangeable ammonium groups shows the same behavior as before with higher release rate in water than in NaCl 1 M. In this case, the soluble fraction is lower when compared to C3-120, as a consequence of the lower content in P(SSAmC$_{16}$-co-GMA20). On the contrary, for the system C4-120 with high P(SSAmC$_{16}$-co-GMA20) content, a significant acceleration of the release rate was found and the selectivity observed between water and NaCl 1 M was vanished. In fact, this system shows the same release rate in both of the solutions used in this study, but the released fraction at 30 days is almost half of the P(SSAmC$_{16}$-co-GMA20) content.

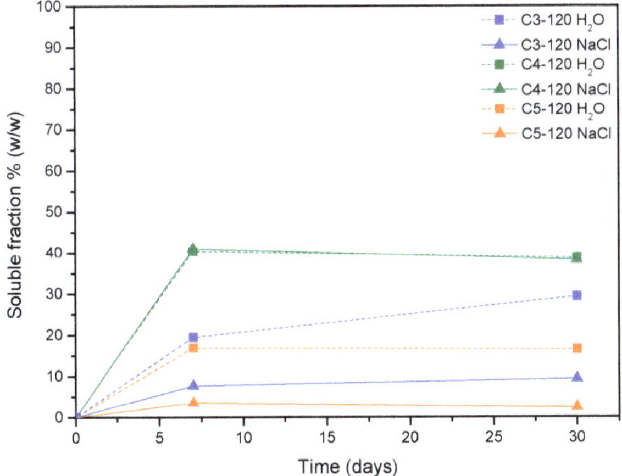

Figure 5. Evolution of soluble fraction % (w/w) after immersion in ultra-pure water and aqueous NaCl 1 M solution of cured C3-120, C4-120, C5-120 polymeric coatings.

The influence of the blend composition is more clearly depicted in Figure 6, where the soluble fraction % (w/w) after seven days testing in ultra-pure water or aqueous NaCl 1 M solution is plotted as a function of the P(VBCHAM-co-AA20) content of the P(SSAmC$_{16}$-co-GMA20)/P(VBCHAM-co-AA20) coatings. The respective TOC and TN values are also plotted, for reasons of comparison. As seen, the trends of all techniques used are in agreement. Moreover, the results show that the released material decreases with the increase of the P(VBCHAM-co-AA20) content, supporting the previous discussions that this material originates from the P(SSAmC$_{16}$-co-GMA20) copolymer.

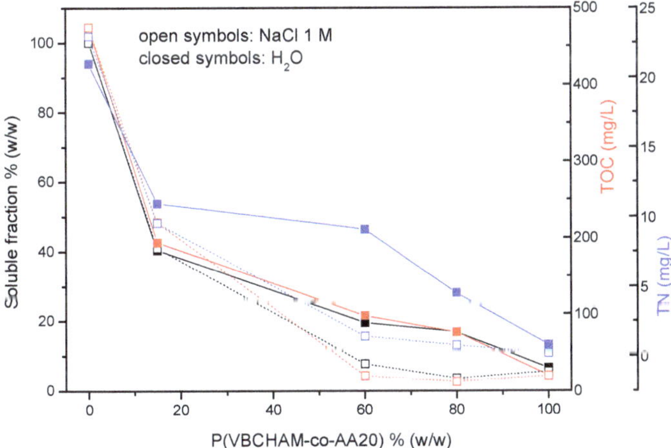

Figure 6. Influence of the P(VBCHAM-co-AA20) % (w/w) content of P(SSAmC$_{16}$-co-GMA20)/P(VBCHAM-co-AA20) 40/60 w/w polymeric coatings on the soluble fraction (%) and the TOC, TN values of the solutions after immersion of the coatings for 7 days in ultra-pure water and aqueous NaCl 1 M solutions. Apart from pure copolymers, all other coatings were cured at 120 °C.

The C/N molar ratio for the C3-C5 coatings, cured at 120 °C after seven days immersion in ultra-pure water or salt solution is shown in Figure 7a as a function of the P(VBCHAM-co-AA20) % (*w/w*) content of the coatings. Similar results are also obtained for 30 days immersion, as shown in Figure 7b. As seen, for the rich in P(VBCHAM-co-AA20) coatings (C3-120 and C5-120), the values of the C/N molar ratio are in the region 16–20, suggesting the release of AmC$_{16}$ cations in these cases. Due to the high content of reactive units of the copolymers, crosslinking reaction is expected to take place in a large extent. Moreover, most P(SSAmC$_{16}$-co-GMA20) chains are expected to be crosslinked for these coatings and release of the whole polymeric chain is not possible. On the other hand, for the C4-120 coating containing P(SSAmC$_{16}$-co-GMA20) in a large excess, it is possible that a part of these chains are not crosslinked. As a consequence, in salt solution they turn to water-soluble and they can be dissolved, leading to the observation of a C/N molar ratio of ~27. Finally, regardless of blend composition, a C/N molar ratio of ~20 is observed for all of the coatings in ultra-pure water, suggesting the release of cetyl trimethylammonium cations, as discussed earlier.

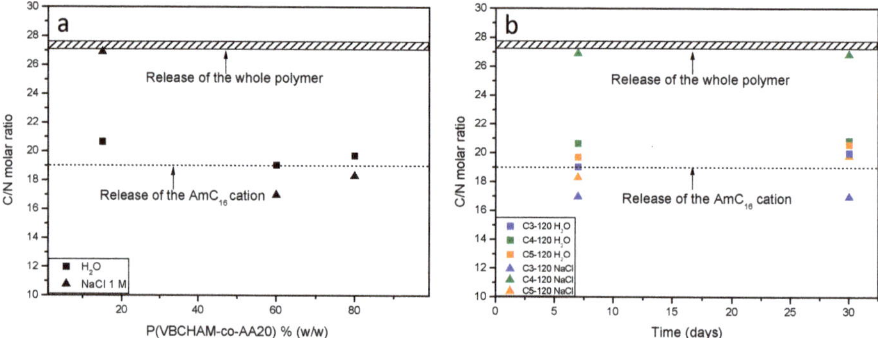

Figure 7. (**a**) Influence of the P(VBCHAM-co-AA20) % (*w/w*) content of the cured P(SSAmC$_{16}$-co-GMA20)/P(VBCHAM-co-AA20) 40/60 *w/w* polymeric coatings on the C/N molar ratio determined from the TOC and TN values of the solutions after immersion of the coatings for seven days in ultra-pure water and aqueous NaCl 1 M solutions, (**b**) Evolution of the C/N molar ratio, determined from the TOC/TN studies, after immersion in ultra-pure water and NaCl 1 M for different time periods of cured C3-120, C4-120, C5-120 polymeric coatings.

Since the salinity of the aqueous solution is a decisive parameter controlling the release levels of the coatings that are developed in the present study, the influence of NaCl concentration was also investigated, using the C3-120 coating. It is reminded that only the AmC$_{16}$ cations may be released from this coating, see Figure 4. The evolution of the soluble fraction in ultra-pure water and aqueous NaCl solutions of varying concentrations (0.25 M, 0.50 M and 1 M) is shown in Figure 8a. Moreover, the variation of the soluble fraction with the NaCl concentration after treatment for 7 days are depicted in Figure 8b, and are compared with the respective TOC variation. It is clear that for NaCl concentrations higher than 0.5 M the release levels are low and do not depend significantly on the salt concentration. However, as the salt concentration decreases below 0.5 M, the release levels progressively increase to attain the maximum release fraction observed in ultra-pure water.

The morphology of the coatings in terms of SEM examination of the self-supported membranes was also studied. Thus, the initial membranes of the C3 system (cured and uncured) together with the respective membranes after treatment for 60 days are depicted in Figure 9. As it is observed, the initial membranes differ depending on their initial thermal treatment. The uncured C3–RT sample shows a particulate structure, while the cured C3-120, due to the higher temperature during thermal treatment, shows a more homogeneous structure. In both cases, the samples after 60 days in aqueous solutions show a more homogeneous structure, as revealed by the cross section images shown in Figure 9. This is mainly due to the swelling of the membrane, as it is seen by the thickness increase from about 20 μm to 40 μm for the C3-120 sample. Thus, the release of the active groups cannot be visualized by SEM, since their competition to swelling vanishes the ability to show any difference.

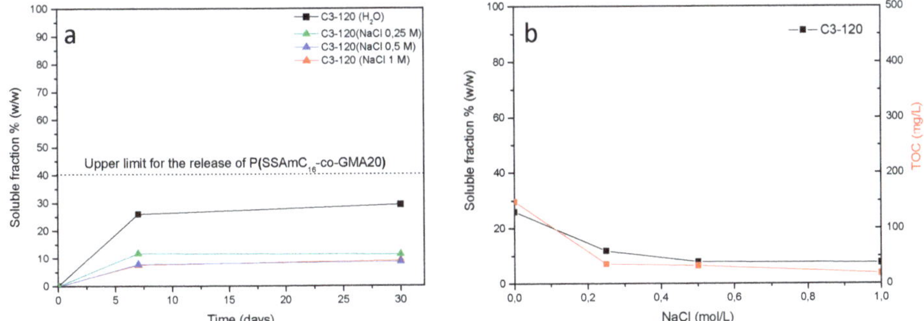

Figure 8. (**a**) Soluble fraction % (w/w) after immersion in ultra-pure water and NaCl 0.25 M, 0.5 M, 1 M for different time periods of cured C3-120 polymeric coatings; (**b**) Soluble fraction % (w/w) with the NaCl concentration after treatment for seven days when compared with the respective TOC variation.

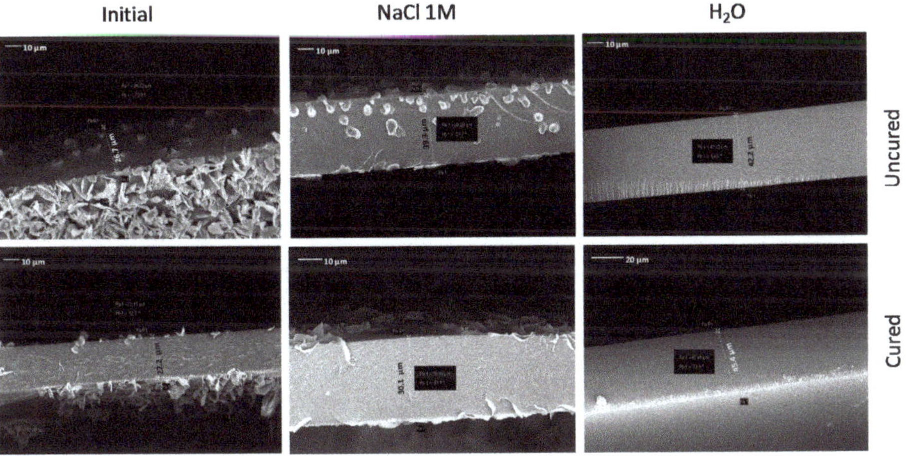

Figure 9. Scanning electron microscopy (SEM) images of uncured C3-RT and cured C3-120 polymeric coatings before and after immersion in ultra-pure water and NaCl 1 M for 60 days.

As a closing remark, it should be noticed that the synthetic protocols of the coatings and characterization protocols concerning the release in aqueous solutions of varying salinity were established in the present work, using covalently attached and releasable hexadecylammonium moieties. Nevertheless, respective units with alternative alkyl chains, for instance, dodecyl chains can also be applied to modulate biocidal activity and its interrelation to the release rate.

4. Conclusions

Different coatings based on covalently and electrostatically bound ammonium groups at various ratios were developed by combination and crosslinking of the respective copolymers bearing the proper active biocidal groups together with reactive acrylic acid and glycidyl methacrylate moieties. The stabilization of the coating was performed by curing at 120 °C, and the release of the active groups was studied for periods up to three months in ultra-pure water and NaCl 1 M solutions in comparison to the uncured analogues. The rate of the release, as well as the type of the releasable species, was monitored by a combination of analytic techniques, like gravimetric and TOC, TN analysis. The ratio of TOC/TN was used as a decisive factor for the type of the releasable species. Depending on the mobile and immobilized ammonium groups' content and the curing temperature, coatings with controllable release behavior were developed. More interestingly, the release rate in NaCl 1 M solutions was lower than the respective in ultra-pure water, mainly because the internal complexation of immobilized oppositely charged species results in the release of AmC_{16} groups, even in ultra-pure water. This phenomenon together with the low release rate in NaCl 1 M solutions makes these biocidal coatings potentially useful candidates both for freshwater and sea water applications.

Supplementary Materials: The following are available online at www.mdpi.com/2079-6412/8/1/8/s1, Figure S1: Evolution of the TOC values of the solutions, after immersion in ultra-pure water and aqueous NaCl 1 M solution for different time periods of: (a) Uncured C1-RT, C2-RT, C3-RT polymeric coatings and (b) Cured C1-120, C2-120, C3-120 polymeric coatings, Figure S2: Evolution of the TN values of the solutions, after immersion in ultra-pure water and aqueous NaCl 1 M solution for different time periods of: (a) Uncured C1-RT, C2-RT, C3-RT polymeric coatings and (b) Cured C1-120, C2-120, C3-120 polymeric coatings, Figure S3: Evolution of the C/N molar ratio, determined from the TOC/TN studies, after immersion in ultra-pure water and NaCl 1 M for different time periods of uncured polymeric coatings $PSSAmC_{16}/P(VBCHAM$-co-$AA20)$ 40/60 w/w.

Acknowledgments: The authors thank Vassilios Dracopoulos from the Institute of Chemical Engineering Sciences (ICE/HT-FORTH) for the SEM characterization.

Author Contributions: Joannis K. Kallitsis and Georgios Bokias conceived and designed the experiments and wrote parts of the manuscript; Vlasoula Bekiari performed the experiments concerning the TOC and TN analysis; Nikos D. Koromilas performed the experiments concerning polymer synthesis and analysis; Denisa Druvari performed the coatings' development and characterization.

Conflicts of Interest: The authors declare no conflict of interest.

References

1. Krumm, C.; Tiller, J.C. Antimicrobial Polymers and Surfaces–Natural Mimics or Surpassing Nature? In *Bio-Inspired Polymers*, 1st ed.; Bruns, N., Kilbinger, A.F.M., Eds.; RSC Polymer Chemistry Series: Cambridge, UK, 2017; pp. 490–522.

2. Yang, Y.; Cai, Z.; Huang, Z.; Tan, X.; Zhan, X. Antimicrobial cationic polymers: From structural design to functional control. *Polym. J.* **2017**, 1–12. [CrossRef]

3. Schnaider, L.; Brahmachari, S.; Schmidt, N.W.; Mensa, B.; Shaham-Niv, S.; Bychenko, D.; Adler-Abramovich, L.; Shimon, L.J.W.; Kolusheva, S.; DeGrado, W.F.; et al. Self-assembling dipeptide antibacterial nanostructures with membrane disrupting activity. *Nat. Commun.* **2017**, *8*, 1365. [CrossRef] [PubMed]

4. Siedenbiedel, F.; Tiller, J.C. Antimicrobial polymers in solution and on surfaces: Overview and functional principles. *Polymers* **2012**, *4*, 46–71. [CrossRef]

5. Kenawy, E.-R.; Worley, S.D.; Broughton, R. The chemistry and applications of antimicrobial polymers: A state-of-the-art review. *Biomacromolecules* **2007**, *8*, 1359–1384. [CrossRef] [PubMed]

6. Muñoz-Bonilla, A.; Fernandez-Garcia, M. The roadmap of antimicrobial polymeric materials in macromolecular nanotechnology. *Eur. Polym. J.* **2015**, *65*, 46–62. [CrossRef]

7. Muñoz-Bonilla, A.; Fernandez-Garcia, M. Polymeric materials with antimicrobial activity. *Prog. Polym. Sci.* **2012**, *37*, 281–339. [CrossRef]

8. Álvarez-Paino, M.; Muñoz-Bonilla, A.; Fernández-García, M. Antimicrobial polymers in the nano-world. *Nanomaterials* **2017**, *7*, 48. [CrossRef] [PubMed]

9. Callow, J.A.; Callow, M.E. Trends in the development of environmentally friendly fouling-resistant marine coatings. *Nat. Commun.* **2011**, *2*, 244. [CrossRef] [PubMed]

10. Jiao, Y.; Niu, L.; Ma, S.; Li, J.; Tay, F.R.; Chen, J. Quaternary ammonium-based biomedical materials: State-of-the-art, toxicological aspects and antimicrobial resistance. *Prog. Polym. Sci.* **2017**, *71*, 53–90. [CrossRef]

11. Simoncic, B.; Tomsic, B. Structures of novel antimicrobial agents for textiles—A review. *Text. Res. J.* **2010**, *80*, 1721–1737. [CrossRef]

12. Ren, W.; Cheng, W.; Wang, G.; Liu, Y. Developments in antimicrobial polymers. *J. Appl. Polym. Sci.* **2017**, *55*, 632–639. [CrossRef]

13. Huang, K.-S.; Yang, C.-H.; Huang, S.-L.; Chen, C.-Y.; Lu, Y.-Y.; Lin, Y.-S. Recent advances in antimicrobial polymers: A mini-review. *Int. J. Mol. Sci.* **2016**, *17*, 1578. [CrossRef] [PubMed]

14. Chen, A.; Peng, H.; Blakey, I.; Whittaker, A.K. Biocidal polymers: A mechanistic overview. *Polym. Rev.* **2017**, *57*, 276–310. [CrossRef]

15. Zubris, D.L.; Minbiole, K.P.; Wuest, W.M. Polymeric quaternary ammonium compounds: Versatile antimicrobial materials. *Curr. Top. Med. Chem.* **2017**, *17*, 305–318. [CrossRef] [PubMed]

16. Xue, Y.; Xiao, H.; Zhang, Y. Antimicrobial polymeric materials with quaternary ammonium and phosphonium salts. *Int. J. Mol. Sci.* **2015**, *16*, 3626–3655. [CrossRef] [PubMed]

17. Ji, W.; Koepsel, R.R.; Murata, H.; Zadan, S.; Campbell, A.S.; Russell, A.J. Bactericidal specificity and resistance profile of poly(quaternary ammonium) polymers and protein–poly(quaternary ammonium) conjugates. *Biomacromolecules* **2017**, *18*, 2583–2593. [CrossRef] [PubMed]

18. Kenawy, E.-R.; Abdel-Hay, F.I.; El-Shanshoury, A.E.-R.R.; El-Newehy, M.H. Biologically active polymers. V. Synthesis and antimicrobial activity of modified poly(glycidyl methacrylate-*co*-2-hydroxyethyl methacrylate) derivatives with quaternary ammonium and phosphonium salts. *J. Polym. Sci. A Polym. Chem.* **2002**, *40*, 2384–2393. [CrossRef]

19. Ganewatta, M.S.; Tang, C. Controlling macromolecular structures towards effective antimicrobial polymers. *Polymer* **2015**, *63*, A1–A29. [CrossRef]

20. Tiwari, A. *Handbook of Antimicrobial Coatings*, 1st ed.; Elsevier Science & Technology Books: New York, NY, USA, 2017; pp. 1–596. ISBN 0128119829.

21. Li, H.; Bao, H.; Bok, K.X.; Lee, C.Y.; Li, B.; Zin, M.T.; Kang, L. High durability and low toxicity antimicrobial coatings fabricated by quaternary ammonium silane copolymers. *Biomater. Sci.* **2016**, *4*, 299–309. [CrossRef] [PubMed]

22. Sisti, L.; Cruciani, L.; Totaro, G.; Vannini, M.; Berti, C.; Aloisio, I.; Di Gioia, D. Antibacterial coatings on poly(fluoroethylenepropylene) films via grafting of 3-hexadecyl-1-vinylimidazolium bromide. *Prog. Org. Coat.* **2012**, *73*, 257–263. [CrossRef]

23. Wei, T.; Tang, Z.; Yu, Q.; Chen, H. Smart antibacterial surfaces with switchable bacteria-killing and bacteria-releasing capabilities. *ACS Appl. Mater. Interfaces* **2017**, *9*, 37511–37523. [CrossRef] [PubMed]

24. Yu, Q.; Wu, Z.; Chen, H. Dual-function antibacterial surfaces for biomedical applications. *Acta Biomater.* **2015**, *16*, 1–13. [CrossRef] [PubMed]

25. Voo, Z.X.; Khan, M.; Xu, Q.; Narayanan, K.; Ng, B.W.J.; Ahmad, R.B.; Hedrick, J.L.; Yang, Y.Y. Antimicrobial coatings against biofilm formation: The unexpected balance between antifouling and bactericidal behavior. *Polym. Chem.* **2016**, *7*, 656–668. [CrossRef]

26. Voo, Z.X.; Khan, M.; Narayanan, K.; Seah, D.; Hedrick, J.L.; Yang, Y.Y. Antimicrobial/antifouling polycarbonate coatings: Role of block copolymer architecture. *Macromolecules* **2015**, *48*, 1055–1064. [CrossRef]

27. Kaura, R.; Liu, S. Antibacterial surface design—Contact kill. *Prog. Surf. Sci.* **2016**, *91*, 136–153. [CrossRef]

28. Alarfaj, A.A.; Lee, H.H.; Munusamy, M.A.; Ling, Q.-D.; Kumar, S.; Chang, Y.; Chen, Y.-M.; Lin, H.-R.; Lu, Y.-T.; Wu, G.-J.; et al. Development of biomaterial surfaces with and without microbial nanosegments. *J. Polym. Eng.* **2016**, *36*, 1–12. [CrossRef]

29. Gao, J.; White, E.M.; Liu, Q.; Locklin, J. Evidence for the phospholipid sponge effect as the biocidal mechanism in surface-bound polyquaternary ammonium coatings with variable cross-linking density. *ACS Appl. Mater. Interfaces* **2017**, *9*, 7745–7751. [CrossRef] [PubMed]

30. Santos, M.R.E.; Fonseca, A.C.; Mendonça, P.V.; Branco, R.; Serra, A.C.; Morais, P.V.; Coelho, J.F.J. Recent developments in antimicrobial polymers: a review. *Materials* **2016**, *9*, 599. [CrossRef] [PubMed]

31. Kougia, E.; Tselepi, M.; Vasilopoulos, G.; Lainioti, G.Ch.; Koromilas, N.D.; Druvari, D.; Bokias, G.; Vantarakis, A.; Kallitsis, J.K. Evaluation of antimicrobial efficiency of new polymers comprised by covalently attached and/or electrostatically bound bacteriostatic species, based on quaternary ammonium compounds. *Molecules* **2015**, *20*, 21313–21327. [CrossRef] [PubMed]

32. Koromilas, N.D.; Lainioti, G.Ch.; Vasilopoulos, G.; Vandarakis, A.; Kallitsis, J.K. Synthesis of antimicrobial block copolymers bearing immobilized bacteriostatic groups. *Polym. Chem.* **2016**, *7*, 3562–3575. [CrossRef]

33. Druvari, D.; Koromilas, N.D.; Lainioti, G.Ch.; Bokias, G.; Vasilopoulos, G.; Vandarakis, A.; Baras, I.; Dourala, N.; Kallitsis, J.K. Polymeric quaternary ammonium-containing coatings with potential dual contact-based and release-based antimicrobial activity. *ACS Appl. Mater. Interfaces* **2016**, *8*, 35593–35605. [CrossRef] [PubMed]

34. Oikonomou, E.K.; Iatridi, Z.; Moschakou, M.; Damigos, P.; Bokias, G.; Kallitsis, J.K. Development of Cu^{2+}-and/or phosphonium-based polymeric biocidal materials and their potential application in antifouling paints. *Prog. Org. Coat.* **2012**, *75*, 190–199. [CrossRef]

35. Thalberg, K.; Lindman, B.; Bergfeldt, K. Phase behavior of systems of polyacrylate and cationic surfactants. *Langmuir* **1991**, *7*, 2893–2898. [CrossRef]

36. Oikonomou, E.; Bokias, G.; Kallitsis, J.K.; Iliopoulos, I. Formation of hybrid wormlike micelles upon mixing cetyl trimethylammonium bromide with poly(methyl methacrylate-*co*-sodium styrene sulfonate) copolymers in aqueous solution. *Langmuir* **2011**, *27*, 5054–5061. [CrossRef] [PubMed]

37. Mathioudakis, G.N.; Soto Beobide, A.; Koromilas, N.D.; Kallitsis, J.K.; Bokias, G.; Voyiatzis, G.A. Evaluation of the release characteristics of covalently attached or electrostatically bound biocidal polymers utilizing SERS and UV-Vis absorption. *eXPRESS Polym. Lett.* **2016**, *10*, 750–761. [CrossRef]

Article

Antimicrobial Films Based on Chitosan and Methylcellulose Containing Natamycin for Active Packaging Applications

Serena Santonicola [1,2], Verónica García Ibarra [1], Raquel Sendón [1], Raffaelina Mercogliano [2] and Ana Rodríguez-Bernaldo de Quirós [1,*]

[1] Department of Analytical Chemistry, Nutrition, and Food Science, Faculty of Pharmacy, University of Santiago de Compostela, 15782 Santiago de Compostela, Spain; serena.santonicola@unina.it (S.S.); veronica.garcia.ibarra@usc.es (V.G.I.); raquel.sendon@usc.es (R.S.)

[2] Department of Veterinary Medicine and Animal Production, University of Naples, 80137 Naples, Italy; raffaella.mercogliano@unina.it

* Correspondence: ana.rodriguez.bernaldo@usc.es; Tel.: +34-881-814-965

Academic Editors: Sami Rtimi and Stefanos Giannakis
Received: 12 September 2017; Accepted: 14 October 2017; Published: 24 October 2017

Abstract: Biodegradable polymers are gaining interest as antimicrobial carriers in active packaging. In the present study, two active films based on chitosan (1.5% w/v) and methylcellulose (3% w/v) enriched with natamycin were prepared by casting. The antimicrobial's release behavior was evaluated by immersion of the films in 95% ethanol (v/v) at different temperatures. The natamycin content in the food simulant was determined by reversed-high performance liquid chromatography with diode-array detection (HPLC-DAD). The apparent diffusion (D_P) and partition ($K_{P/S}$) coefficients were calculated using a mathematical model based on Fick's Second Law. Results showed that the release of natamycin from chitosan based film ($D_P = 3.61 \times 10^{-13}$ cm^2/s) was slower, when compared with methylcellulose film ($D_P = 3.20 \times 10^{-8}$ cm^2/s) at the same temperature ($p < 0.05$). To evaluate the antimicrobial efficiency of active films, cheese samples were completely covered with the films, stored at 20 °C for 7 days, and then analyzed for moulds and yeasts. Microbiological analyses showed a significant reduction in yeasts and moulds (7.91 log CFU/g) in samples treated with chitosan active films ($p < 0.05$). The good compatibility of natamycin with chitosan, the low D_P, and antimicrobial properties suggested that the film could be favorably used in antimicrobial packagings.

Keywords: active packaging; chitosan; methylcellulose; natamycin

---

1. Introduction

Antimicrobial food packaging is one of the most promising applications of active packaging, and acts to reduce, inhibit or retard microorganism growth that could contaminate the packaged food [1–3].

During recent decades, due to the environmental and economic implications of the use of materials made from petrochemical derivatives, various research groups have been looking toward green polymers as alternatives to film and coating manufacturing for plastic packages [4–9]. Bio-based packaging materials can be created on the basis of polymers directly extracted/removed from natural materials [10]. Elaboration of these films and coatings has been possible thanks to the filmogenic capacity of natural biopolymers, which have a good aptitude for forming a continuous and cohesive matrix with adequate mechanical properties [10,11]. Chitosan is a natural weak cationic polysaccharide, derived from deacetylation of chitin, which is the major component of the shells of crustaceans such as crab, shrimp, and crawfish [2,12]. It is non-toxic, biodegradable, biofunctional, and biocompatible [13,14]. Chitosan-based films have good mechanical properties, and has some advantages over other biomolecule-based polymers used as packaging materials due to its antibacterial behavior [13–16]. Methylcellulose application as a film and coating component is also very attractive. It is one of the most important commercial cellulose ethers, and it has been used in many industrial applications [17]. Methylcellulose films have a flexible and transparent character. They also possess low ox18ygen and moisture vapor transmission rates when compared to other hydrophilic edible films [18].

Biodegradable packaging materials can act as supporters of antimicrobials [19]. Much research has been devoted to the design of antimicrobial packaging containing natural antimicrobial agents for specific or broad microbial inhibition [20]. Bacteriocins are an attractive option, as they constitute natural preservatives, avoiding the addition of synthetic compounds to food [3,21]. These antimicrobial proteins/peptides, produced by bacteria, are nontoxic and nonantigenic to humans, and have GRAS (generally recognized as safe) status [3,22]. Natamycin, produced during fermentation by the bacterium *Streptomyces natalensis*, is a naturally occurring antifungal agent classified as a macrolide polyene, which acts through the specific interaction with ergosterol of yeast membranes [23,24]. According to Directive 95/2/EC, natamycin may be used for the surface treatment of semi-hard and semi-soft cheese and dry cured sausage at a maximum level of 1 mg/dm^2 in the outer 5 mm of the surface [25,26]. Natamycin has no adverse effect on the rind or the flavor of the cheese. The depth of penetration of this antimicrobial compound depends on of the initial concentration, cheese type and storage time [27]. Incorporation of antimicrobials in food interfaces by the use of films helps to decrease the rate of diffusion from the surface to the bulk of the product assuring the maintenance of high concentrations of the active agent where it is required [28]. In addition, due to the low water solubility of natamycin, the incorporation into a coating improves distribution in the cheese and, therefore, the surface protection from mould growth [4]. This antimicrobial agent has been successfully used in different active systems [29–31]. Chitosan coating containing natamycin decreased mold/ yeast population on Saloio cheese after 27 days of storage [4]. Moreover, the application of chitosan films determined an increase of the shelf life of different cheese types, such as Mozzarella [32], Emmental [33], Regional Saloio [34], and Apulia spreadable cheese [35]. Furthermore, natamycin-impregnated cellulose-based films showed inhibitory effects against Penicillium *roquefortii* on the surface of Gorgonzola cheese [36] and, in combination with nisin, prolonged the shelf life of sliced mozzarella cheese by 6 days compared to the control [37]. Methylcellulose and wheat gluten films containing natamycin showed ability in the prevention and control of toxigenic moulds on dairy products [18].

For the selection of an antimicrobial, the possible interactions among the antimicrobial, the film-forming biopolymer, and other food components, which can modify the antimicrobial activity and the characteristics of the film, must be considered [15]. Diffusion (*D*) and partition (*K*) coefficients for antimicrobials in packaging films can help to design efficient active packaging and to predict the shelf-life of food products [19,38]. Therefore, the aim of this study was to develop two active films, based on chitosan and methylcellulose, incorporating natamycin as antimicrobial agent, and to study the release behavior of antimicrobial agents from the active films. The antimicrobial effectiveness of chitosan and methylcellulose films containing natamycin to prevent yeast and mould growth on cheese surface was also investigated.

2. Materials and Methods

2.1. Materials

The materials used to prepare the active films were: Chitosan with a degree of deacetylation of approximately 75–85% (medium molecular weight, 200–800 mPa·s viscosity, soluble in 1% Acetic acid aqueous solution) (Sigma chemicals, St-Louis, MO, USA); Glycerol (molecular biology grade, Calbiochem); Methylcellulose (3500–6000 mPa viscosity) (Sigma-Aldrich, Darmstadt, Germany). Natamycin was provided by Sigma (Steinheim, Germany). Ethanol (analytical grade), Methanol (HPLC grade), Acetonitrile (HPLC grade) were provided by Merck (Darmstadt, Germany). Acetic acid solution

(HPLC grade) was provided by Sigma-Aldrich (Germany). The water used to prepare all solutions was purified by a Milli-Q water purification system (Millipore) (Bedford, MA, USA).

2.2. Preparation of Films

Chitosan films (1.5% w/v) were prepared by dissolving chitosan in acetic acid aqueous solution 1% (v/v). Subsequently, the solution, containing glycerol as plasticizer (0.2 g/g biopolymer), was kept under stirring for 2 h at a constant temperature at 80 °C and then 12 h at room temperature until the chitosan was fully dissolved.

Methylcellulose (3% w/v) was mixed with a water-ethanol solution (50:50 v/v) and then homogenized for 5 min. After the addition of glycerol (0.4 g/g biopolymer), the solution was kept under stirring, and heated to 80 °C for 2 h. Natamycin was incorporated at room temperature into the film solutions to reach the final concentration of 0.01% (w/v). The solutions were then cast in 8.5 cm polyacrylic plates and dried at 30 °C for 12 h. A saturated solution of magnesium nitrate was put in the oven to achieve a relative humidity of 53%.

2.3. Film Thickness Measurement

The thickness of the samples was determined using a manual digital micrometer (0.001 mm, Mitutoyo, Mizonokuchi, Japan). Measurements were repeated in 5 different regions of each sample and then an average value was calculated.

2.4. Experimental Procedure for Kinetics of Natamycin Release

During migration tests, the films were fixed in glass tubes so that both sides of the tested films were in contact with food simulant. According to EU Commission Regulation No 10/2011 [39], ethanol 95% (v/v) was used as a substitute food simulant for fatty food. Chitosan and methylcellulose films were cut into pieces of 14.7 cm^2 area and immersed into 20 mL of ethanol 95% (v/v). The migration kinetics of the antimicrobial agent from methylcellulose film were studied at different temperatures: 10, 20 and 40 °C ($\pm$0.2 °C). To compare the release behavior between the different active films, natamycin migration test from chitosan film was performed at 40 °C ($\pm$0.2 °C).

To determine the amount of natamycin released, aliquots (500 µL) of the food simulant were taken out from the tubes at preset times, filtered and injected by HPLC-DAD (high-performance liquid chromatographic with diode-array detection). The analytical conditions were obtained by modification of a previously reported method [40]. The preserving agent released from the film into the simulant was determined as follows: the calculation of the final migration level included the correction for changing simulant volume, as well as for the amount of natamycin taken during the previous sampling. Then, to calculate the remaining amount in the film, a piece of film was introduced in a glass tube containing 20 mL of ethanol 80% (v/v) solution at 40 °C for 24 h. The experiment was performed in duplicate.

HPLC-DAD Analysis

An HPLC HP1100 system (Hewlett Packard, Waldbronn, Germany) equipped with a quaternary pump, a degassing device, an autosampler, a column thermostating system, a diode-array detector (DAD), and Agilent Chem-Station for LC and LC/MS systems software (Agilent, Santa Clara, CA, USA), was used. Separation was performed on a Kromasil ODS (C18) (150 $\times$ 3.20 mm^2 i.d., 5 mm particle size) column thermostatted at 25 °C. Acetonitrile (A) and Milli-Q water (B) were used as mobile phase. The injection volume was 20 µL. Samples were eluted in gradient mode under the following conditions: 0 min (20% A–80% B); 10 min (60% A–40% B); 15 min (60% A–40% B); 20 min (20% A–80% B). The flow rate was 0.6 mL/min. Three selected wavelengths were set in DAD detector, 291, 304 and 319 nm, corresponding to the three absorption peaks of the characteristic natamycin spectrum [40]. The wavelength used to quantify the antifungal was 304 nm.

2.5. Diffusion Coefficient (D) and Partition Coefficient (K) Measurement

The diffusion coefficients of natamycin from the active films into the substitute food simulant ethanol 95% (*v*/*v*) were calculated using a mathematical model based on Fick's Second Law (1):

$$\frac{\partial C_p}{\partial t} = D \frac{\partial^2 C_p}{\partial x^2} \tag{1}$$

where C_p is the concentration of the migrant in the film at time t and position x.

An analytical solution of this differential equation, that describes the diffusion kinetics, was proposed by Crank [41]. After a slight modification, this can be expressed by the following Equations (2) and (3) [42]:

$$\frac{m_{F,t}}{A} = c_{P,0}\rho_P d_P \left(\frac{\alpha}{1+\alpha}\right) \times \left[1 - \sum_{n=1}^{\alpha} \frac{2\alpha(1+\alpha)}{1+\alpha+\alpha^2 q_n^2} \exp\left(-D_P t \frac{q_n^2}{d_P^2}\right)\right] \tag{2}$$

$$\alpha = \frac{1}{K_{P/F}} \frac{V_F}{V_P} \tag{3}$$

where $m_{F,t}$ is the mass of the migrant transferred from P into F after time t, (µg); A is the area of P in contact with F (cm^2); $C_{P,0}$ is the initial concentration of the migrant in P (mg/kg); ρ_P is the density of P (g/cm^3); t is the migration time (s); d_p is the thickness of P (cm); V_P is the volume of P (cm^3); V_F is the volume of F (cm^3); q_n is the positive root of the equation tan $q_n = -\alpha \cdot q_n$; D_P is the diffusion coefficient for the migrant in the polymer (cm^2/s); $K_{P/F}$ is the partition coefficient for the migrant between P and F.

Partition coefficient between the film and food simulant ($K_{P/S}$) was calculated according to the following Equation (4):

$$K_{P/S} = \frac{C_P}{C_S} \tag{4}$$

where C_P is the concentration of a substance in the film at equilibrium, in µg/g; C_S is the concentration of a substance in the simulant at equilibrium, in µg/g.

Experimental data were fitted to the proposed model using Solver function of the commercial software Microsoft Excel 2007® (Redmond, WA, USA).

To measure the fit between the experimental and estimated data the root of mean-square error % (RMSE (%)) was calculated according to the following Equation (5):

$$\text{RMSE}(\%) = \frac{1}{M_{P,0}} \sqrt{\frac{1}{n}\sum_{i=1}^{n} \left((m_{F,t})_{\exp,i} - (m_{F,t})_{pred,i}\right)^2} \times 100 \tag{5}$$

where n is the number of experimental points per migration/release curve; i is the number of observations; $M_{P,0}$ is the initial amount of the migrant in the polymer (µg).

2.6. Microbiological Analysis

A commercial semi-hard cheese was purchased from a local supermarket and stored in a refrigerator (4 °C) until use. Fifteen pieces of cheese were randomly assigned to five treatments: samples coated with Polyethylene (used as blank), samples coated with chitosan and with methylcellulose films, and finally samples coated with chitosan and with methylcellulose films containing natamycin.

Cheese samples (10 g) were completely covered with the films (28.26 cm^2), sealed in polyethylene bags, and finally stored at 20 °C for 7 days.

At the times of bacterial enumeration, cheese samples were aseptically removed from their packaging and the films were separated from cheese slices with sterile forceps. An aliquot (10 g) of cheese was aseptically collected and placed in 400 mL homogenizing bag along with 90 mL of 0.1% (*w*/*v*) of peptone water and massaged for 60 s at high speed in a Stomacher (AES Chemunex, Coburg, FR, Germany). Decimal dilutions were prepared from the initial homogenate. After that, 0.1 mL were spread onto potato dextrose agar (PDA) plates and incubated at 25 °C for 3 days, before counting colonies. Three samples of each treatment were analyzed.

2.7. Statistical Analysis

Statistical analysis of data was performed with package SPSS 15.0 (SPSS Inc., Chicago, IL, USA). ANOVA test was applied to determine significant differences ($p < 0.05$) according to the release rate of antimicrobial agent from methylcellulose films at different temperatures, and from chitosan and methylcellulose film at the same temperature, according to mould/yeast counts.

3. Results and Discussion

3.1. Film Appearance Characterization

All of the films were transparent. The average thickness ranged between 38 and 51 µm for chitosan, and from 56 to 76 µm for methylcellulose films.

Natamycin was added to the film solutions at concentrations of 6.6 and 3.3 mg per gram of chitosan and methylcellulose, respectively. The increase in natamycin concentration (from 0.01% to 0.1% and 1%), affected the appearance of both types of films, making them opaque and unsuitable for use. This behavior has also been observed by other authors [24]. For the migration kinetics, it is essential to determine the initial concentration of the active compound in the developed films. The concentrations of natamycin in the chitosan and methylcellulose films prior to the migration assays were 0.6% (*w*/*w*) and 0.3% (*w*/*w*), respectively.

3.2. Natamycin Release Kinetics

Figures 1 and 2 illustrate the release profiles of natamycin from chitosan and methylcellulose films in the substitute food simulant (Ethanol 95% (*v*/*v*)) at different storage temperatures as a function of time.

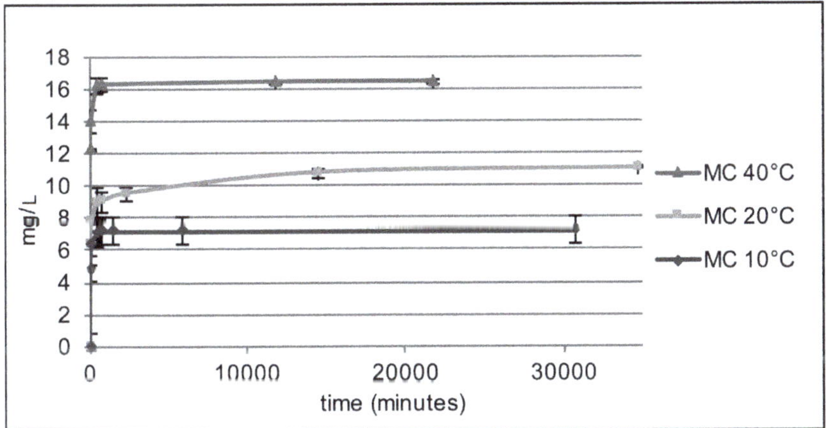

Figure 1. Release profiles of natamycin from methylcellulose films (MC) at different temperatures. Each dot represents the mean of the experimental data with an error bar of 2 replications.

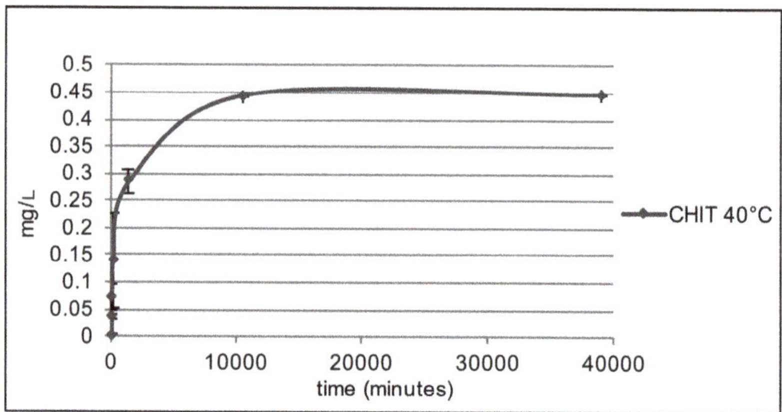

Figure 2. Release profiles of natamycin from chitosan film (CHIT) at 40 °C. Each dot represents the mean of the experimental data with an error bar of 2 replications.

With respect to the migration kinetics, results showed that the release of natamycin from the chitosan-based films was slower, when compared with the methylcellulose films at the same temperature ($p < 0.05$). In addition, after 12 h at 40 °C, methylcellulose film released about 90% of the natamycin, while less than 15% of the initial amount of natamycin in the chitosan film was released when equilibrium conditions were reached.

The diffusion and the transport mechanism of the active agents from the film matrix to the food surfaces are the most important factors in developing an antimicrobial food packaging. The diffusion coefficient indicates the rate at which the release of the active agent takes place. Diffusivities can be used to quantify the release behavior of the antimicrobials, and also to obtain information about polymeric networks [11]. The partition and apparent diffusion coefficients of natamycin from chitosan (at 40 °C) and methylcellulose films (at 10, 20, 40 °C) are included in Table 1. For methylcellulose films, in order to test the linearity between the D_p and the temperature, an Arrhenius-type equation was applied, and $R^2 = 0.9384$ was obtained.

Table 1. Partition ($K_{P/S}$) and apparent diffusion (D_P) coefficients (mean values) of natamycin from the active films in food simulant at different temperatures (T).

Films	T	D_P (cm^2/s)	$K_{P/S}$	RMSE (%)
chitosan + natamycin	40 °C	3.61×10^{-13}	1054.11	0.63
methylcellulose + natamycin	10 °C	7.30×10^{-10}	271.64	5.38
methylcellulose + natamycin	20 °C	1.14×10^{-9}	127.85	4.59
methylcellulose + natamycin	40 °C	3.20×10^{-8}	3.74	10.93

According to the literature [19], for each tested temperature, and for the same temperature (40 °C), the diffusion coefficients of natamycin from methylcellulose films were higher, when compared to chitosan films. As expected, the diffusion process of natamycin from methylcellulose films generally occurred faster at higher temperatures ($p < 0.05$). However, literature data showed that the temperature variations could result in a negligible change in the diffusion coefficient of antimicrobial compound from chitosan film [19].

The low release of natamycin from chitosan film could be related to different factors. Apparent diffusion coefficients for natamycin from alginate/chitosan film into water at an order of magnitude in the range of 10^{-11}–10^{-12} have been reported [11]. These values are considered very low when compared with diffusivities of other antimicrobials incorporated in polymeric matrices, suggesting a chemical interaction between natamycin and chitosan [11]. Chen et al. [43] attributed this fact to a possible electrostatic interaction between NH_3^+ groups of chitosan and COO^- groups of antimicrobials. Methylcellulose carries no electrical charge [44], and the possibility of interactions between the polymer and the antimicrobial is lower.

Diffusion coefficients at an order of magnitude of 10^{-10} and 10^{-12} cm^2/s have been reported for natamycin from chitosan films to phosphate-buffered saline solution and cheese, respectively [4]. The diffusion coefficient was higher for the phosphate-buffered saline solution because of the swelling effect in the release phenomenon [4]. Similar results were obtained by Hanušova et al. [45] with polyvinyl dichloride lacquer coatings also used as a natamycin carrier in cheese packaging. Water-ethanol solution 95% (v/v), used as substitute food simulant in our study, exhibits lower water activity, which subsequently results in lower degradation, swelling and solubilization phenomena in biopolymers, making the antimicrobial diffusion through the film matrix difficult [38]. Moreover, chitosan maintains its structure in a neutral environment, but is solubilized and degraded in an acidic medium [46]. As the water–ethanol solution used in the present study had a neutral pH, chitosan film demonstrated higher stability with a compact structure and, therefore, lower diffusion coefficient values [19]. A major advantage of slow release over direct addition of the antimicrobial into the food is continuous microbial inhibition obtained over an extended period [47]; in contrast, a rapid release may cause migration of the active agent to internal parts of the food, reducing the protection at the surface [1].

The partition coefficient ($K_{P/S}$) indicates the ratio between the concentration of the active compound in the film and the concentration in the food simulant at equilibrium [48,49]. The partition coefficients, shown in Table 1, correspond to the values predicted by the mathematical model used. Higher $K_{P/S}$ values are achieved with higher concentrations of the active agent in the film; in contrast, a lower $K_{P/S}$ indicates that more migrant is absorbed into food from the polymer. However, various parameters, such as temperature, pH, the chemical structure of the migrant, molecular size and structure, and the fat content of foods, can influence the partition coefficient [49].

In order to measure the fit between the experimental and estimated data, the root of mean-square error % (RMSE (%)) was calculated. Generally, acceptable values were obtained. In particular, the best fit between the experimental and estimated data was found for chitosan films [8].

3.3. Antimicrobial Activity of Films

The microbial assays were performed under accelerated conditions at 20 °C. The mean values of the counts obtained for moulds and yeasts were 7.91 log (CFU/g) and 8.25 log (CFU/g), respectively, for chitosan and methylcellulose films containing natamycin. Values slightly higher for mould/yeast counts were observed in cheese coated with chitosan (8.30 log CFU/g), methylcellulose (8.95 log CFU/g) and polyethylene (8.27 log CFU/g) films, when compared to the active films. It is important to note that results were obtained under no-inoculation conditions. Therefore, large differences in the counts were not to be expected for the whole cheese, and not only on the cheese surface [4].

Antimicrobial agents can be applied by dipping, spraying, or brushing to food surfaces for controlling microbial growth [18]. However, these techniques are laborious, and have limited benefits, because of a rapid loss of activity resulting from the interaction between the antimicrobial compound and food components, and from a dilution phenomenon occurring when the additive diffuses to the bulk of the food [24,29]. The use of packaging films containing antimicrobial agents could better control the migration of the agents away from the surface [50]. In earlier studies, antimycotic activity of natamycin-incorporating films has been shown against several moulds [30–32]. The low antimicrobial activity of methylcellulose films could be explained by the rapid release, which may cause migration of the active agent to internal parts of the food, reducing the protection at the surface [1]. On the contrary, the slow release of natamycin from chitosan film into the food surface determined a significant ($p < 0.05$) reduction of mould/yeast counts with respect to polyethylene films, which were used as a blank. Chitosan has been tested as a carrier of other natural antimicrobials, e.g., lysozyme [51], lysozyme and EDTA [52], essential oils [13,38], and nisin [19]. In particular, several studies have reported that chitosan antimicrobial films containing natamycin possess the potential ability to inhibit microorganisms on food products [4,11,24]. The incorporation of natamycin in chitosan-based film could act as an additional post-processing safety measure, once the inhibitory effect on microbial growth of both components [4].

Antimicrobial films or coatings resulted in more effective microorganism inhibition when applied to nutrient media than when applied to real systems, owing to the complex structure of foods [18]. In fact, the minimum inhibitory concentration of natamycin against *A. niger* and *P. roquefortii* was found to be two times higher in cheese application with respect to in vitro studies [18].

4. Conclusions

Briefly, the kinetics results showed that, at the same temperature, the release of natamycin from chitosan-based film ($D_P = 3.61 \times 10^{-13}$ cm^2/s) was slower, when compared with methylcellulose film ($D_P = 3.20 \times 10^{-8}$ cm^2/s) ($p < 0.05$). Moreover, a significant reduction in yeast and mould was observed in cheese samples treated with chitosan films containing natamycin ($p < 0.05$). The controlled release of natamycin from chitosan film would address the limitation of rapid loss of natamycin applied directly to the cheese surface. However, further studies are needed to measure their antimicrobial activities on selected microorganisms and on other real food surfaces.

Acknowledgments: The study was financially supported by the "Consellería de Cultura, Educación e Ordenación Universitaria, Xunta de Galicia", Ref. No. GRC 2014/012. V. García Ibarra was supported by a graduate fellowship from the Secretaría Nacional de Educación Superior, Ciencia, Tecnología e Innovación of Ecuador.

Author Contributions: Ana Rodríguez Bernaldo de Quirós, Raquel Sendón and Raffaelina Mercogliano conceived and designed the experiments and analyzed the data; Serena Santonicola performed the experiments and wrote the paper; and Veronica García Ibarra performed the microbiological experiments.

Conflicts of Interest: The authors declare no conflict of interest.

References

1. Appendini, P.; Hotchkiss, J.H. Review of antimicrobial food packaging. *Innov. Food Sci. Emerg.* **2002**, *3*, 113–126. [CrossRef]

2. da Silva, M.A.; Iamanaka, B.T.; Taniwaki, M.H.; Kieckbusch, T.G. Evaluation of the antimicrobial potential of alginate and alginate/chitosan films containing potassium sorbate and natamycin. *Packag. Technol. Sci.* **2013**, *26*, 479–492. [CrossRef]

3. Santiago-Silva, P.; Soares, N.F.F.; Nóbrega, J.E.; Júnior, M.A.W.; Barbosa, K.B.F.; Volp, A.C.P.; Zerdas, E.R.M.A.; Würlitzer, N.J. Antimicrobial efficiency of film incorporated with pediocin (ALTA_ 2351) on preservation of sliced ham. *Food Control* **2009**, *20*, 85–89. [CrossRef]

4. Fajardo, P.; Martins, J.T.; Fuciños, C.; Pastrana, L.; Teixeira, J.A.; Vicente, A.A. Evaluation of a chitosan-based edible film as carrier of natamycin to improve the storability of Saloio cheese. *J. Food Eng.* **2010**, *101*, 349–356. [CrossRef]

5. Imran, M.; Revol-Junelles, A.M.; Martyn, A.; Tehrany, E.A.; Jacquot, M.; Desobry, S. Active food packaging evolution: Transformation from micro- to nanotechnology. *Crit. Rev. Food Sci. Nutr.* **2010**, *50*, 799–821. [CrossRef] [PubMed]

6. Jamshidian, M.; Tehrany, E.A.; Imran, M.; Jacquot, M.; Desobry, S. Poly-lactic acid: Production, applications, nanocomposites, and release studies. *Compr. Rev. Food Sci.* **2010**, *9*, 552–571. [CrossRef]

7. Mastromatteo, M.; Mastromatteo, M.; Conte, A.; Del Nobile, M.A. Advances in controlled release devices for food packaging applications. *Trends Food Sci. Technol.* **2010**, *21*, 591–598. [CrossRef]

8. Rodríguez-Martínez, A.V.; Sendón, R.; Abad, M.J.; Gonzalez-Rodríguez, M.V.; Barros-Velazquez, J.; Aubourg, S.P.; Paseiro-Losada, P.; Rodríguez-Bernaldo de Quirós, A. Migration kinetics of sorbic acid from polylactic acid and seaweed based films into food stimulants. *LWT Food Sci. Technol.* **2016**, *65*, 630–636. [CrossRef]

9.   Siracusa, V.; Rocculi, P.; Romani, S.; Dalla Rosa, M. Biodegradable polymers for food packaging: A review. *Trends Food Sci. Technol.* **2008**, *19*, 634–643. [CrossRef]

10.   Bourtoom, T. Edible protein films: Properties enhancement. *Int. Food Res. J.* **2009**, *16*, 1–9.

11.   Da Silva, M.A.; Bierhalz, A.C.K.; Kieckbusch, T.G. Modelling natamycin release from alginate/chitosan active films. *Int. J. Food Sci. Technol.* **2012**, *47*, 740–746. [CrossRef]

12.   Lago, M.A.; Sendón, R.; Rodríguez-Bernaldo de Quirós, A.; Sanches-Silva, A.; Costa, H.S.; Sánchez-Machado, D.I.; Soto Valdez, H.; Angulo, I.; Aurrekoetxea, G.P.; Torrieri, E.; et al. Preparation and characterization of antimicrobial films based on chitosan for active food packaging applications. *Food Bioprocess Technol.* **2014**, *7*, 2932–2941. [CrossRef]

13.   Abdollahi, M.; Rezaei, M.; Farzi, G. A novel active bionanocomposite film incorporating rosemary essential oil and nanoclay into chitosan. *J. Food Eng.* **2012**, *111*, 343–350. [CrossRef]

14.   Aider, M. Chitosan application for active bio-based films production and potential in the food industry: Review. *LWT Food Sci. Technol.* **2010**, *43*, 837–842. [CrossRef]

15.   Campos, C.A.; Gerschenson, L.N.; Flores, S.K. Development of edible films and coatings with antimicrobial activity. *Food Bioprocess Technol.* **2011**, *4*, 849–875. [CrossRef]

16.   No, H.K.; Meyers, S.P.; Prinyawiwatkul, W.; Xu, Z. Applications of chitosan for improvement of quality and shelf life of foods: A review. *J. Food Sci.* **2007**, *72*, 87–100. [CrossRef] [PubMed]

17.   Nasatto, P.L.; Pignon, F.; Silveira, J.L.M.; Duarte, M.E.R.; Noseda, M.D.; Rinaudo, M. Methylcellulose, a cellulose derivative with original physical properties and extended applications. *Polymers* **2015**, *7*, 777–803. [CrossRef]

18.   Ture, H.; Eroglu, E.; Ozen, B.; Soyer, F. Effect of biopolymers containing natamycin against *Aspergillus niger* and *Penicillium roquefortii* on fresh kashar cheese. *Int. J. Food Sci. Technol.* **2011**, *46*, 154–160. [CrossRef]

19.   Imran, M.; Klouj, A.; Revol-Junelles, A.M.; Desobry, S. Controlled release of nisin from HPMC, sodium caseinate, poly-lactic acid and chitosan for active packaging applications. *J. Food Eng.* **2014**, *143*, 178–185. [CrossRef]

20.   Dainelli, D.; Gontard, N.; Spyropoulosc, D.; Zondervan-van den Beukend, E.; Tobback, P. Active and intelligent food packaging: Legal aspects and safety concerns. *Trends Food Sci. Technol.* **2008**, *19*, S103–S112. [CrossRef]

21.   Cotter, P.D.; Hill, C.; Ross, R.P. Bacteriocins: Developing innate immunity for food. *Nat. Rev. Microbiol.* **2005**, *3*, 777–788. [CrossRef] [PubMed]

22.   Halami, P.M.; Chandrashekar, A. Enhanced production of pediocin C20 by anative strain of Pediococcus acidilatici C20 in an optimized food-grade medium. *Process Biochem.* **2005**, *40*, 1835–1840. [CrossRef]

23.   Ce´, N.; Norena, C.P.Z.; Brandelli, A. Antimicrobial activity of chitosan films containing nisin, peptide P34, and natamycin. *CyTA-J. Food* **2012**, *10*, 21–26. [CrossRef]

24.   Ollé Resa, C.; Gerschenson, L.; Jagus, R. Effect of natamycin on physical properties of starch edible films and their effect on Saccharomyces cerevisiae activity. *Food Bioprocess Technol.* **2013**, *6*, 3124–3133. [CrossRef]

25.   EU Parliament and Council Directive No 95/2/EC of 20 February 1995 on Food Additives Other Than Colours and Sweeteners. Available online: http://eur-lex.europa.eu/legal-content/EN/TXT/?uri=LEGISSUM%3Al21070a (accessed on 24 October 2017).

26.   EFSA Panel on Food Additives and Nutrient Sources added to Food (ANS). Scientific Opinion on the use of natamycin (E 235) as a food additive. *EFSA J.* **2009**, *7*, 1412.

27.   Davidson, P.M.; Sofos, J.N.; Branen, A.L. *Antimicrobials in Food*, 3rd ed., CRC Press. Boca Raton, FL, USA, 2005.

28.   Kristo, E.; Koutsoumanis, K.; Biliaderis, C. Thermal, mechanical and water vapor barrier properties of sodium caseinate films containing antimicrobials and their inhibitory action on Listeria monocytogenes. *Food Hydrocoll.* **2008**, *22*, 373–386. [CrossRef]

29.   Cong, F.; Zhang, Y.; Dong, W. Use of surface coatings with natamycin to improve the storability of Hami melon at ambient temperature. *Postharvest Biol. Technol.* **2007**, *46*, 71–75. [CrossRef]

30.   Reps, A.; Drychowski, L.J.; Tomasik, J.; Winiewska, K. Natamycin in ripening cheeses. *Pak. J. Nutr.* **2002**, *1*, 243–247.

31.   Var, I.; Erginkaya, Z.; Güven, M.; Kabak, B. Effects of antifungal agent and packaging material on microflora of Kashar cheese during storage period. *Food Control* **2006**, *17*, 132–136. [CrossRef]

32.   Altieri, C.; Scrocco, C.; Sinigaglia, M.; Del Nobile, M.A. Use of chitosan to prolong mozzarella cheese shelf life. *J. Dairy Sci.* **2005**, *88*, 2683–2688. [CrossRef]

33.   Coma, V.; Martial-Gros, A.; Garreau, S.; Copinet, A.; Salin, F.; Deschamps, A. Edible antimicrobial films based on chitosan matrix. *J. Food Sci.* **2002**, *67*, 1162–1169. [CrossRef]

34. Cerqueira, M.A.; Lima, A.M.; Souza, B.W.S.; Teixeira, J.A.; Moreira, R.A.; Vicente, A.A. Functional polysaccharides as edible coatings for cheese. *J. Agric. Food Chem.* **2009**, *57*, 1456–1462. [CrossRef] [PubMed]
35. Gammariello, D.; Chillo, S.; Mastromatteo, M.; Di Giulio, S.; Attanasio, M.; Del Nobile, M.A. Effect of chitosan on the rheological and sensorial characteristics of Apulia spreadable cheese. *J. Dairy Sci.* **2008**, *91*, 4155–4163. [CrossRef] [PubMed]
36. De Oliveira, T.M.; de Fátima Ferreira Soares, N.; Pereira, R.M.; de Freitas Fraga, K. Development and evaluation of antimicrobial natamycin-incorporated film in Gorgonzola Cheese conservation. *Packag. Technol. Sci.* **2007**, *20*, 147–153. [CrossRef]
37. Dos Santos Pires, A.C.; De Ferreira Soares, N.F.; De Andrade, N.J.; Mendes Da Silva, L.H.; Peruch Camilloto, G.; Campos Bernardes, P. Development and evaluation of active packaging for sliced mozzarella preservation. *Packag. Technol. Sci.* **2008**, *21*, 375–383. [CrossRef]
38. Sánchez-González, L.; Cháfer, M.; González-Martínez, C.; Chiralt, A.; Desobry, S. Study of the release of limonene present in chitosan films enriched with bergamot oil in food stimulants. *J. Food Eng.* **2011**, *105*, 138–143. [CrossRef]
39. Commission Regulation (EU) No 10/2011 on Plastic Materials and Articles Intended to Come into Contact with Food. Available online: https://www.fsai.ie/uploadedFiles/Reg10_2011.pdf (accessed on 12 September 2017).
40. Paseiro-Cerrato, R.; Otero-Pazos, P.; Rodríguez-Bernaldo de Quirós, A.; Sendón, R.; Angulo, I.; Paseiro-Losada, P. Rapid method to determine natamycin by HPLC-DAD in food samples for compliance with EU food legislation. *Food Control* **2013**, *33*, 262–267. [CrossRef]
41. Crank, J. *The Mathematics of Diffusion*, 2nd ed.; Oxford University Press Inc.: New York, NY, USA, 1975; pp. 69–88.
42. Simoneau, C. Applicability of Generally Recognised Diffusion Models for the Estimation of Specific Migration in Support of EU Directive 2002/72/EC. Available online: http://ibebvi.studiorauw.eu/src/Frontend/Files/Labo/5/files/guideline%20modelling_70a.pdf (accessed on 24 October 2017).
43. Chen, M.C.; Yeh, G.H.C.; Chiang, B.H. Antimicrobial and physicochemical properties of methylcellulose and chitosan films containing a preservative. *J. Food Process. Preserv.* **1996**, *20*, 379–390. [CrossRef]
44. Nuijts, R.M.; Nuijts, R.M.M.A. *Ocular Toxicity of Intraoperatively Used Drugs and Solutions*; Kugler Publications: New Amsterdam, The Netherlands, 1995; pp. 19–20.
45. Hanušova, K.; Štastná, M.; Votavová, L.; Klaudisová, K.; Dobiáš, J.; Voldrich, M. Polymer films releasing nisin and/or natamycin from polyvinyldichloride lacquer coating: Nisin and natamycin migration, efficiency in cheese packaging. *J. Food Eng.* **2010**, *99*, 491–496. [CrossRef]
46. Agnihotri, S.A.; Mallikarjuna, N.N.; Aminabhavi, T.M. Recent advances on chitosan-based micro- and nanoparticles in drug delivery. *J. Control. Release* **2004**, *100*, 5–28. [CrossRef] [PubMed]
47. Chung, D.; Chikindas, M.; Yan, K. Inhibition of Saccharomyces cerevisiae by slow release of propyl paraben from a polymer coating. *J. Food Protect.* **2001**, *64*, 1420–1424. [CrossRef]
48. Silva, A.S.; Freire, J.M.C.; García, R.S.; Franz, R.; Losada, P.P. Time-temperature study of the kinetics of migration of dpbd from plastics into chocolate, chocolate spread and margarine. *Food Res. Int.* **2007**, *40*, 679–686. [CrossRef]
49. Tehrany, E.A.; Desobry, S. Partition coefficients in food/packaging systems: A review. *Food Addit. Contam.* **2004**, *21*, 1186–1202. [CrossRef] [PubMed]
50. Ouattara, B.; Simard, R.E.; Piett, G.; Begin, A.; Holley, R.A. Inhibition of surface spoilage bacteria in processed meats by application of antimicrobial films prepared with chitosan. *Int. J. Food Microbiol.* **2000**, *62*, 139–148. [CrossRef]
51. Duan, J.; Park, S.I.; Daeschel, M.A.; Zhao, Y. Antimicrobial chitosan-lysozyme films and coatings for enhancing microbial safety of mozzarella cheese. *J. Food Sci.* **2007**, *72*, 359–362. [CrossRef] [PubMed]
52. Del Nobile, M.T.; Gammariello, D.; Conte, A.; Attanasio, M. A combination of chitosan, coating and modified atmosphere packaging for prolonging Fior di latte cheese shelf life. *Carbohyd. Polym.* **2009**, *78*, 151–156. [CrossRef]

Review

Fused Deposition Modelling as a Potential Tool for Antimicrobial Dialysis Catheters Manufacturing: New Trends vs. Conventional Approaches

Essyrose Mathew , Juan Domínguez-Robles, Eneko Larrañeta *and Dimitrios A. Lamprou *

School of Pharmacy, Queen's University Belfast, Belfast BT9 7BL, UK
* Correspondence: e.larraneta@qub.ac.uk (E.L.); d.lamprou@qub.ac.uk (D.A.L.); Tel.: +44-28-9097-2360 (E.L.); +44-28-9097-2617 (D.A.L.)

Received: 13 July 2019; Accepted: 10 August 2019; Published: 14 August 2019

Abstract: The rising rate of individuals with chronic kidney disease (CKD) and ineffective treatment methods for catheter-associated infections in dialysis patients has led to the need for a novel approach to the manufacturing of catheters. The current process requires moulding, which is time consuming, and coated catheters used currently increase the risk of bacterial resistance, toxicity, and added expense. Three-dimensional (3D) printing has gained a lot of attention in recent years and offers the opportunity to rapidly manufacture catheters, matched to patients through imaging and at a lower cost. Fused deposition modelling (FDM) in particular allows thermoplastic polymers to be printed into the desired devices from a model made using computer aided design (CAD). Limitations to FDM include the small range of thermoplastic polymers that are compatible with this form of printing and the high degradation temperature required for drugs to be extruded with the polymer. Hot-melt extrusion (HME) allows the potential for antimicrobial drugs to be added to the polymer to create catheters with antimicrobial activity, therefore being able to overcome the issue of increased rates of infection. This review will cover the area of dialysis and catheter-related infections, current manufacturing processes of catheters and methods to prevent infection, limitations of current processes of catheter manufacture, future directions into the manufacture of catheters, and how drugs can be incorporated into the polymers to help prevent infection.

Keywords: 3D printing; catheters; dialysis; extrusion; infections; manufacturing

---

1. Introduction

In the United Kingdom, there are currently around 30,000 people on dialysis [1]. Over the last 50 years, the provision of chronic dialysis has steadily increased with over 2 million people worldwide being treated with dialysis [2]. Dialysis is a procedure used to remove waste products and excess fluid from the blood when there is decreased kidney function. With over 3000 kidney transplants taking place every year in the United Kingdom and 5000 on the waiting list, Chronic Kidney Disease (CKD) is a pressing issue [3]. The National Health Service (NHS) England estimated spending of approximately £1.45 billion on CKD in 2009–2010 [4]. CKD is present in about 65% of people over 85 years of age [5]. CKD is estimated to affect 11–13% of the population and is forecast to become the fifth leading cause of death worldwide by 2040 [6].

After the age of 18, nephrons in the kidney decline by around 7000 per year. Nephrons cannot be regenerated by the body. Renal blood flow also declines after the age of 40 and there is an increased vascular resistance, so the level of blood that reaches glomeruli for filtration is reduced [5]. Therefore, an aging population may also be a factor in the increasing number of CKD patients as renal function declines with age. The two most common types of dialysis include Haemodialysis (HD) and peritoneal dialysis (PD). In HD, by using a central venous catheter (CVC), the catheter is inserted into a large vein usually in the chest. The catheter is made up of two lumens, one in which the blood is taken out of the body and filtered by an external machine. Filtered blood is then returned through the other lumen. In PD, the inside lining of the abdomen is used as a filter, where a catheter is placed in the abdomen, through which fluid is pumped, and as blood passes through vessels lining the peritoneal cavity, waste products and excess fluid are drawn out of blood and into dialysis fluid [7].

There are mainly two types of dialysis catheters (Figure 1A). A tunnelled catheter is one that is tunnelled under the skin to a separate exit site; it is preferred for long-term use due to increased stability with most of the catheter being in the body. Non-tunnelled catheters (NTHCs) are inserted into the body with the majority of the catheter present outside the body. NTHCs are used mainly for temporary vascular access. The type of catheter that is used in patients can also have an effect on the rate of infection. Tunnelled dialysis catheters are often used in patients with end-stage kidney disease and used as a longer-term vascular access route. Tunnelled catheters have a lower risk of infection as they have subcutaneous tunnels that increase the distance between the bloodstream and skin [8]. NTHC is used when urgent vascular access is required and is usually used for short-term vascular access. However, due to an increased risk of complications, NTHCs are the least preferred form of vascular access for chronic HD patients. On a recent study, done on a cohort of 154 patients receiving renal replacement therapy with acute kidney disease, patients with tunnelled dialysis catheters had significantly better delivery of the therapy compared to those with the NTHCs [9]. There was better blood flow and a significantly lower number of complications with tunnelled catheters [10]. A comparable study showed that there was no difference in the first occurrence of infection for tunnelled and NTHCs. However, tunnelled catheters were removed less often [11]. One of the major concerns with the treatment of CKD through HD and PD is the risk of infection.

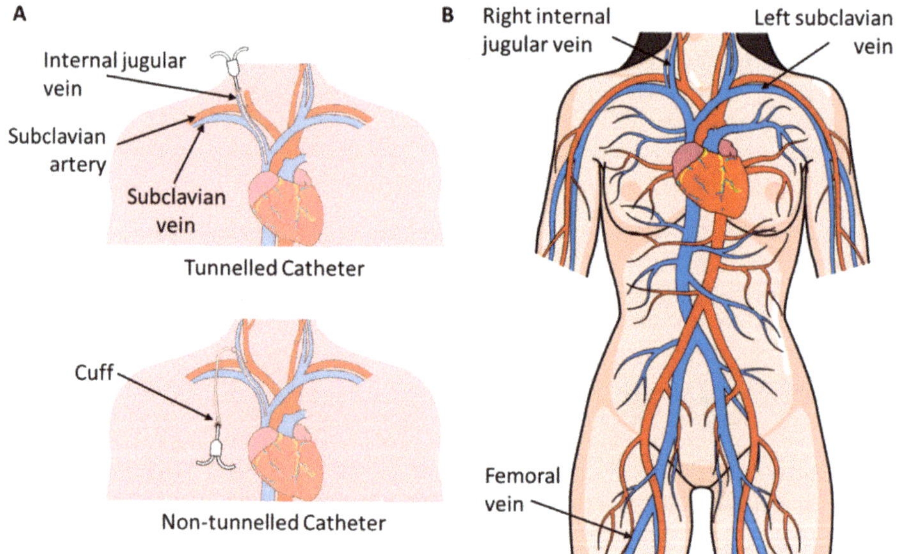

Figure 1. (**A**) Tunnelled catheter and non-tunnelled catheter. (**B**) Catheter Insertion sites.

1.1. Infections

In patients who undergo dialysis, infections after insertion of the catheter is a prominent issue. Bacteraemia can occur in patients between 0.6 to 6.5 episodes per 1000 catheter days [12] and 87.3% of catheter-related infections are caused by Gram-positive bacteria, such as *Staphylococcus aureus*; however, Gram-negative microorganisms, such as *Escherichia coli*, can also be the cause of bacteraemia in patients [13]. Catheter infection complications can occur in 15–40% of cases. These are most often present in infections caused by *S. aureus* [14]. In a study done on patients with end-stage kidney disease, 12.8% of the population developed *Staphylococcus aureus* bacteraemia (SAB) [15]. There was not a significant difference in the risk of SAB between cuffed and non-cuffed catheters, with the risk being higher in patients with central venous catheters (CVCs) than peritoneal catheters [15].

Bacteria that are antibiotic-resistant, which can increase the difficulty of treatment, often cause catheter-associated infections. The cost to the NHS in the United Kingdom of methicillin-resistant *Staphylococcus aureus* (MRSA) in HD patients is estimated at £1.4 million [4]. According to the same report, annual costs per patient are £24,043 for HD and £20,078 for PD.

The risk of infection can also rise according to the position in which the catheter is inserted. The risk of infection is higher for the subclavian site (chest) of insertion and lower for femoral (groin) and jugular (neck) (Figure 1B). Insertion success is also significantly higher for femoral and jugular sites compared to subclavian [16]. However, a controversial study has stated that there is no statistically significant relationship between infection and insertion site of the device [17].

Around 64% of hospital-acquired infections are caused by viable bacteria attaching to medical devices and implants [18]. Catheter-related infections are typically distinguished from colonisation—tip culture yielding >103 CFU [19]. Therefore, it is important to find a solution that prevents bacteria from attaching to the surface of medical devices, such as catheters with antimicrobial properties that prevent bacteria from being able to proliferate. Biofilm formation is characteristic of around 80% of all human infection. Bacterial biofilms are protected by a matrix of polymeric substances on their surface and enable multidrug resistance to occur within these matrices [20]. Figure 2A shows an example of a peritoneal dialysis catheter-related infection. A factor that contributes to the high risk of infection in dialysis patients is fibrin sheath formation (Figure 2B).

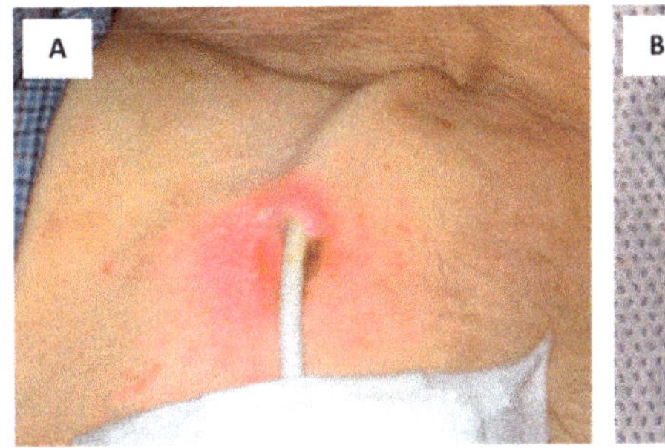
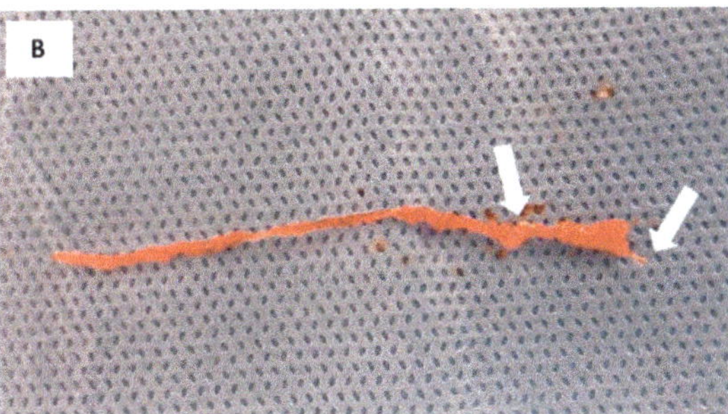

Figure 2. (**A**) Example of peritoneal dialysis catheter exit site infection. (**B**) Fibrin sheath formation was detected around the tip of the removed catheter (arrow). Reproduced with permission from: Lok et al. [21] and Mogi et al. [22]. Copyright 2018 Elsevier.

1.2. Fibrin Sheath Formation

Fibrin sheath formation is a complication that occurs with HD catheters. It is common in cases of late catheter dysfunction [23]. A fibrin sleeve can form around the catheter, which can affect the function of the catheter and cause ineffective HD. Fibrin sheath formations can occur from 14 days of catheter insertion, according to animal studies [24]. At the earlier stages of fibrin formation, the sheath may be non-occlusive but can cause occlusions at later stages [25]. One method that has been used to prevent fibrin sheath formation is the use of water infused surface protection (WISP) on CVCs. WISP creates a boundary layer on the inner lumen of CVCs, so that blood does not come into contact with the lumen walls. This method showed a significant reduction in the surface density of adhering proteins and therefore had a negative effect on fibrin sheath formation [12]. May et al. has also shown the addition of Sharklet micropatterns onto the surface of the catheter to have 86% and 80% reduction in platelet adhesion and fibrin sheath formation, respectively. In addition to the reduction in fibrin sheath formation, this method also reduced bacterial adhesion of *S. aureus* and *Staphylococcus epidermidis* [26].

2. Current Processes

Current processes for the manufacture of catheters require moulding. Molten polymer is poured into a rubber mould and cured. The curing process is time consuming, as the heat needs to flow from outside the catheters through the entire body by thermal conduction or by infrared radiation (IR) [27]. Injection moulding has also been used for the manufacture of catheters, and is the process by which molten polymer is injected into a mould [27].

Catheters are made with silicone, polyurethane (PU), or latex. Latex allergy is common among individuals so pre-manufactured catheters may create added complications to patient treatment if allergy occurs. A study done on ventricular catheters by Weisenberg et al. stated that ventricular catheters currently are made from a silicone material and are available as straight tubes, which can then be cut appropriately for the dedicated use, in angular configurations or a set length [28]. Thermoplastic elastomers are often used due to their elasticity, which allows the catheters to be inserted with more ease and lower risk of damaging blood vessels. Due to inertness in the body, flexible properties, and blood compatibility, PU has been identified as a good polymer for use in medical devices [29].

Rough edges can encourage bacterial adhesion as well as increased risk of damage to vessels; therefore, smooth surfaces on catheters are preferred. Hydrophilic coatings are sometimes used on catheters as they provide more lubrication, so a lower insertion force is required and there will be lower friction on insertion. A method used currently to help prevent infection occurring in patients with dialysis catheters is the use of coatings around the catheters.

2.1. Catheter Coatings

Coatings are used in catheter manufacturing to support the reduction of infection. These coatings can consist of antimicrobial agents that can help prevent adherence of bacteria onto the surface of catheters. Although coatings have been proven to prevent adherence of bacteria in most cases, coatings have some limitations. Currently depending on the type of coating, a concentration of 2% (*w*/*w*) is needed for an antimicrobial effect, whereas a study with 3D-printed catheter tips has shown antimicrobial properties present from as little as 1% (*w*/*w*) [30].

2.1.1. Pyrogallol Coating

Pyrogallol (PG) coating on a catheter works effectively against *S. aureus*, but much higher concentrations are required in order to work effectively against *E. coli*. However, as these are some of the most common pathogens causing catheter-related infections, it is important to use an antimicrobial agent that is effective against a wide range of pathogens [17]. PG-coated catheters are dependent on concentration for an antimicrobial effect, with a concentration of 125 μg/mL required to have an antimicrobial effect against both *S. aureus* and *E. coli* [31]. PG can be used as an antibiotic free coating and so reduce the potential for antibiotic resistance to occur. Balne et al. used PG- and metal ion-coated catheters with the coating having similar properties to the non-coated catheter with added wettability [31]. The coated catheters also showed significant activity against MRSA strains, which are common bacterial strains associated with catheter infection. PG was tested at concentrations of 0.1%, 1%, and 2% (*w*/*v*), with higher concentrations having a greater zone of inhibition. The antimicrobial properties of PG and PG with antimicrobial metal ions were proven to have broad-spectrum activity [31].

2.1.2. Heparin Coating

Heparin-coated catheters have been shown to decrease fibrin deposits that may increase biofilm formation and are the form of catheter used currently in hospitals to minimise infection. Antimicrobial coated catheters are associated with a lower rate of colonisation and catheter-related infection [19]. Animal studies have shown a decreased rate in thrombus formation with heparin-coated catheters. As HD catheters may remain inserted for several months, heparin has the potential to cause adverse effects [32]. As heparin is an anticoagulant, it can cause bleeding, allergic reactions, and increase the risk of osteoporosis with long-term use. However, it is often used in heparin-bonded catheters to prolong the usefulness of a catheter [33]. Heparin-induced thrombocytopenia (HIT) can occur when patients are exposed to any level of heparin [34]. However, in a study on 130,000 patients with heparin-bonded grafts, the incidence rate of HIT was <0.1%. This is because HIT occurs after systematic administration of heparin. Therefore, in the case of intravenous catheters, there would be a higher risk of developing HIT and so it is a greater cause for concern when administering HD and PD catheters [35].

2.1.3. Silver Particles

Silver particles have antimicrobial activity against both Gram-positive and Gram-negative bacteria. Additionally, this material shows low cytotoxicity. Therefore, silver particles have the potential to be included in catheter coatings [36]. Figure 3 shows a schematic representation of the known mechanisms of antibacterial action of silver nanoparticles, which are: (1) The silver nanoparticles adhere to the bacterial surface; (2) DNA within the bacterial cell is damaged due to the silver nanoparticles; (3) Ag^+ ions are released, which have antimicrobial properties. These ions interact with the proteins in the bacterial cell wall, causing the cell wall to lose functionality; (4) Ag^+ ions disrupt the proton electrochemical gradient in bacteria, resulting in reduced ATP synthesis, which can lead to cell death.

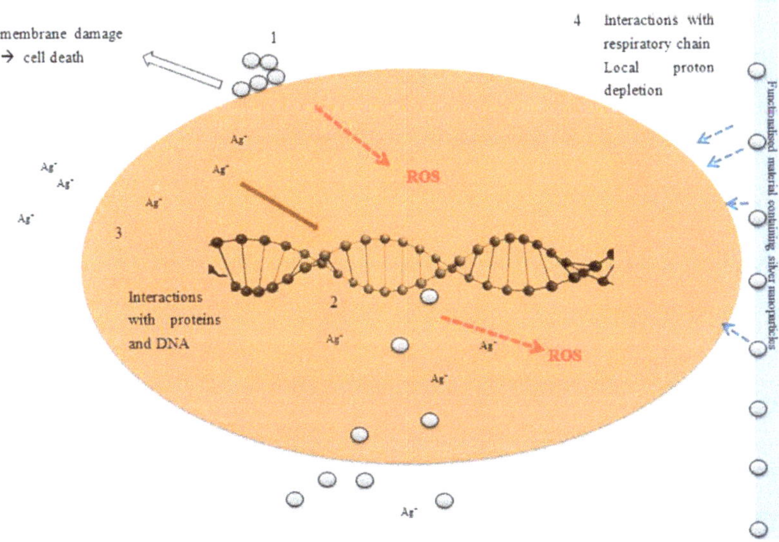

Figure 3. Schematic representation of the known mechanism(s) of antibacterial action of silver nanoparticles and released ionic silver. The numbers 1–4 correspond to the mechanisms described in the paragraphs above. Grey circles indicate silver nanoparticles (NPs) and Ag^+ implies ionic silver released from the NPs. Reproduced with permission from Reidy et al. [37].

The antimicrobial activity is dependent on the dose of silver nanoparticles [38]. Nonetheless, higher doses have been associated with increased cytotoxicity [38]. Interestingly, Kuehl et al. showed that silver coatings had limited activity against *S. aureus* with combination therapy of silver alongside an antibiotic such as vancomycin, producing greater activity against MRSA [39].

Freitas et al. used the sol–gel method for deposition of titanium onto the surface of central venous catheters, followed by the addition of silver particles by irradiation. The results showed that this method did not produce a homogenous coating on the catheters. Further antimicrobials tests also showed that these catheters did not have antimicrobial activity [40]. Additionally, a meta-analysis showed that silver impregnated catheters, which have been used to reduce the risk of infection, were not associated with reduced rates of bacterial colonization or catheter-related bloodstream infections [41]. Therefore, there are still some questions around whether silver nanoparticles have significant antimicrobial activity when incorporated into coatings. Although there are many articles showing the antimicrobial activity of silver nanoparticles, not all of these are tested in vivo [39,40,42]. Additionally, microorganisms vary in their sensitivity to silver [43]. A recent study showed that silver nanoparticles had antimicrobial activity against *E. coli*, *Klebsiella pneumonia*, *Acinetobacter* sp., and *Pseudomonas aeruginosa* at concentrations of 0.5–2.5 µg/µL. However, there was no significant activity against *S. aureus*, i.e., a Gram-positive bacterium, and, as discussed above, infections are most commonly caused by Gram-positive bacteria [42]. Another recent work also confirmed that silver nanoparticles were less effective in *S. aureus* than in *E. coli* [44]. This may be due to the presence of the thicker peptidoglycan layer in Gram-positive bacteria, which may provide some added protection to the bacteria in comparison to Gram-negative bacteria [45]. Various studies have also shown that there is a cause for concern with the increased use of silver nanoparticles in medical devices, as silver resistance strains of bacteria may form [46–48]. Silver nanoparticles have also been shown to create histopathologic abnormalities in the liver, spleen, and lung, as well as toxicity in the muscle. However, there are few studies on the potential toxicities of silver nanoparticles and range of doses that may cause toxicity [49]. Due to these limitations, silver particles are not frequently used in current catheter coating processes.

2.2. Limitations of Coated Catheters

There are limitations to the use of coated catheters. If used over a long period, there is potential for the antimicrobial agent to cause toxicity [32]. A large amount of antimicrobial agent is also required to have a significant antimicrobial effect. The volume of antimicrobial agent required to coat catheters before providing a significant antimicrobial effect is high in comparison to additive manufacturing (AM) techniques.

A major problem with coated catheters is the development of resistance to antimicrobials with catheter-associated infections [31]. Over time, as the antimicrobial agent is released from the coating, the coating becomes thinner, resulting in a lower rate of release of drug. This change in rate of release can increase the potential for antibiotic resistance. Slow drug release from catheter coatings presents the issue of finding the balance between a sufficient amount of drugs in the coating for antimicrobial effect and limiting the coating thickness to ensure physical properties of the catheter remain [50].

Catheter coating also requires the modification of both the inner and outer surfaces of the catheter for optimal effect against bacteria, which lengthens the manufacturing process. Bacterial growth on the inner surface of catheters also requires a higher concentration of antibiotics to stop growth than on the outer surface [51].

It is important to note that the current manufacturing process is generic and is not adapted to the patient's needs. Moreover, it is expensive to create drug-coated polymers as well as there being an increased wastage of material during the coating process. Due to these limitations, it is obvious that there is a need to develop new technologies for catheter manufacture.

3. Future Directions for Catheter Manufacture: Challenges and Expected Impact

3.1. Additive Manufacturing

One of the novel technologies of AM includes 3D printing. This technique is a process in which a digital file made using computer-aided design (CAD) can be printed into a physical object. This allows for a representation of the model to be created using computer software before its final release, which reduces the time for developmental stages of manufacture [52]. As the pharmaceutical and medical devices industries are moving towards personalized medicine and medical devices, the future direction of AM will enable the use of imaging of patients, to design patient-matched devices through computer modelling software and 3D print these devices. 3D Printing is an AM technique in which polymers can be extruded and deposited in multiple layers to create an object. Fused deposition modelling (FDM), to the best of our knowledge, is the only AM technology with published data in the area of catheters and with a promising future in this area. However, use of other 3D printing technologies (e.g., Selective laser sintering (SLS)) might be possible in the near future. The three main principles through which FDM works is the extrusion of polymer, depositing material in successive layers, and cooling of material on the printer bed to solidify structure.

FDM could be used in the manufacturing of catheters, allowing flexibility in the choice of polymer-used design specifications, according to the patient. Table 1 outlines the advantages and disadvantages of FDM.

Table 1. Table summarising advantages and disadvantages of fused deposition modelling.

Advantages	Disadvantages
Rapid Manufacture	Drug needs to have similar or higher melting point to polymer
Less expensive	Small range of thermoplastic polymers
On-demand Fabrication	Lower resolution than stereolithography
Patient Matched Device	Difficult to produce small diameter filament

Risk of infection would persist; therefore, antimicrobial properties are required to reduce infection rates. FDM allows the potentials of antimicrobial filaments to be used in which the polymer has an antimicrobial drug mixed-in. One way in which antimicrobial filaments can be created is through hot-melt extrusion (HME).

The use of additive manufacturing for catheter preparation has been barely described in the literature. There are two papers describing the use of this technology for catheter preparation [30,53]. These basic studies are proof of concept studies and accordingly contain many aspects that should be improved before this technology can be applied to patients. The catheters described in this study are prepared using poly(lactic acid) (PLA), methotrexate, and gentamicin (Figure 4). PLA is not the most appropriate candidate for catheter manufacturing for several reasons that will be discussed further below. Moreover, the resulting catheters showed high surface roughness (Figure 4A,B). This factor should be improved as surface irregularities promote bacterial adhesion and biofilm accumulation [54]. However, there are simple ways to reduce surface roughness for FDM-produced devices [55,56]. Finally, the resulting catheters showed gentamicin release over a period of up to five days. However, the authors did not check if the catheters retained its antimicrobial activity after five days. This is another important point that needs to be considered for future works.

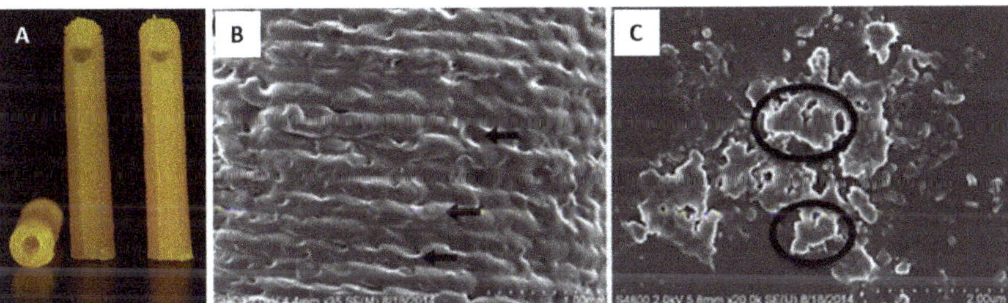

Figure 4. Photograph of the methotrexate-laden three-dimensional (3D)-printed catheter (**A**). Scanning electron microscope images of gentamicin-laden 3D-printed catheters (**B,C**). Multiple amorphous defects seen at 35× magnification suggest gentamicin incorporation into the catheter structure (**B**, arrows). This is confirmed at 20,000× magnification, which highlights the amorphous configuration of gentamicin (**C**, circles). Reproduced with permission from Weisman et al. [30]. Copyright 2019 Elsevier.

3.2. Hot-Melt Extrusion

HME is a process in which heat and pressure are applied to create molten material and force it through an orifice to create uniform filaments. Figure 5 shows a medical tubing extrusion line. Traditional methods such as spray drying (spraying equipment, such as spraying drying, is also being used for coating stents and catheters) involve the use of organic solvents, creating disposal and environmental issues. As HME is a solvent-free process, it is a preferred method due to the reduced waste of organic solvent and its environmentally friendly nature [57]. It is a process in which an active drug can be processed with a polymer as a carrier [58]. A recent study done on wound dressings made from polymer extrudates with antimicrobial metal ions showed the potential for creating patient-specific devices with HME materials [59]. HME extrusion has been used to create filaments for FDM 3D printing. A study by Melocchi et al. uses HME to produce filaments using ethyl cellulose (Eudragit RL), polye(thylene oxide) (PEO), and poly(vinyl alcohol) (PVOH), which were then successfully 3D printed into discs using an FDM printer [52]. HME has also been used to improve the compatibility of certain polymers for FDM printing. For example, pure Eudragit is too brittle to be printed; however, when combined with a plasticizer using HME, filaments are produced with more desirable properties for FDM printing [60]. Alhijjaj et al. also explored the creating altered drug release rates from FDM-printed objects by altering the polymer blends through HME. PEG, PEO, and Tween 80 with Eudragit or Soluplus were studied, as well as blends with PVA, which is commonly used in FDM printing [61]. This method of HME can be used to combine a wide range of polymers to produce filaments with differing mechanical properties to suit the application [62].

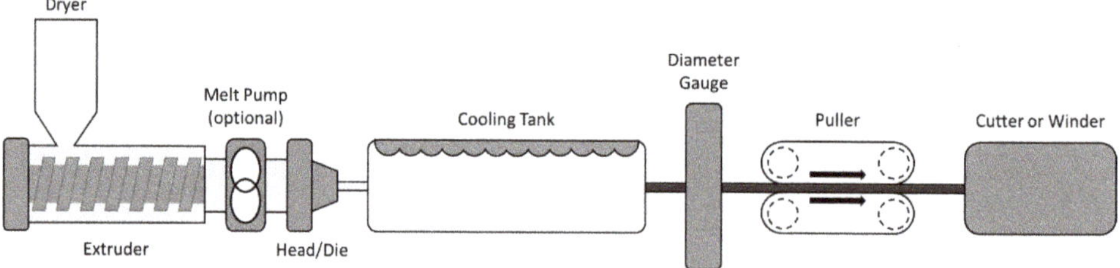

Figure 5. Typical medical tubing extrusion line. This type of extrusion line contains several pieces of equipment, including a drying system, an extruder, a die, a cooling tank, a take-up device (puller), and a winder or cutter.

4. Suitable Materials for Additive Manufacturing

As described above, there are not many reports on the use of AM for catheter manufacturing. However, this promising technology can be a good alternative to prepare medical materials on demand, including catheters. Several materials can be used for this purpose. These materials are mainly polymers that have been used before for 3D printing applications and have been proven to be biocompatible.

4.1. Poly(Lactic Acid)

Poly(lactic acid) (PLA) (Figure 6) is a polymer that is biodegradable and bioresorbable [63,64]. PLA is an inexpensive polymer and easily accessible, that melts in the range of 180–220 °C, making it a suitable polymer for 3D printing. The products of degradation from PLA are easily excreted out of the body by kidneys and products of degradation are non-toxic to the body [65]. Considering that renal function is compromised in CKD patients, the use of biodegradable polymers is not ideal. Moreover, the surface of the catheter will be eroded over time. Accordingly, these types of catheters will not be recommended for prolonged use. Moreover, as PLA can be hydrolysed, it is important to consider the storage of PLA in humid environments as the moisture from the air could hydrolyse the polymer [66]. In biomedical applications, PLA has been used for tissue engineering, sutures, and prostheses [67].

PLA is also generally recognised as a safe material by the food and drug administration (FDA) [68]. Potential drawbacks of PLA include its poor thermal stability and brittleness, which make it less favourable for large-scale manufacturing. However, it is a popular material used in AM and has been proven to be effective in FDM [56]. A study on microfluidic devices has shown the use of PLA through FDM to manufacture medical devices, with less than 1% variability shown between replicate prints [69]. PLA stents have also been produced through FDM, with a printing temperature of 220 °C [70]. This shows that PLA can be successfully used in the process of FDM. PLA has also been proven for its uses within HME, with studies showing the manufacture of dexamethasone implants for the controlled release of immunosuppressive and anti-inflammatory drugs [71,72]. Moreover, PLA has been recently used to manufacture 3D-printed catheters containing a chemotherapeutic agent (methotrexate) or an antibiotic (gentamicin sulphate) to be used as a type of personalized medicine in interventional radiology [30].

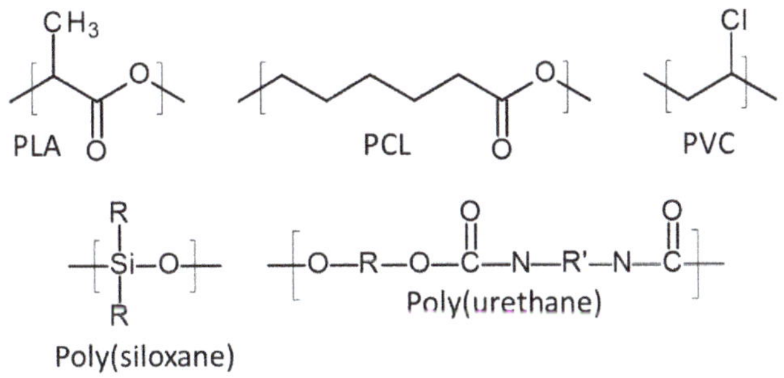

Figure 6. Chemical structures of poly(lactic acid) (PLA), poly(caprolactone) (PCL), poly(vinyl chloride) (PVC), poly(siloxane), and poly(urethane).

4.2. Poly(Caprolactone)

Poly(caprolactone) (PCL) (Figure 6) has a low melting point of ca. 60 °C, which is good for extrusion [63]. PCL is biodegradable and thermally stable so it can withstand the high temperatures used in FDM and HME. PCL also has a low glass transition temperature at −60 °C, making it a more flexible material [73], and has been used to print stents through FDM effectively at a temperature of 220 °C [70]. The stents produced achieved 85–90% accuracy through the FDM printing process. Additionally, Fu et al. produced a progesterone-loaded filament through HME at 190 °C, which was then printed into vaginal rings using FDM at a printing temperature of 195 °C [74].

4.3. Poly(Vinyl Chloride)

Poly(vinyl chloride) (PVC) (Figure 6) is a thermoplastic material with a melting point of around 240 °C. When PVC is plasticised, it can have some advantageous properties, such as good biocompatibility, flexibility, and strength, and has been used in biomedical applications for catheters, gloves, and blood bags [75]. Sharma et al. has shown the effective extrusion of PVC granules into filaments, and further processed the PVC filaments through FDM [76]. This study also tested the mechanical properties of the PVC-printed constructs, highlighting the potential use in HME and FDM. The first plastic catheter manufactured in 1945 using PVC and PU, but nowadays, these catheters are in disuse due to their rigidity/stiffness, thrombogenic capacity, and for causing bacterial adherence. In general, plasticisers can leach from the polymer matrix, as they are not covalently bonded to the polymer. There is also a risk of drugs in the body migrating into plastics, which could lead to drugs falling below the therapeutic threshold [77].

4.4. Thermoplastic Poly(Urethane)

Polyurethanes (PU) are a family of polymers that are widely used in the manufacture of biomedical devices (Figure 6). Thermoplastic polyurethane (TPU) is a polymer commonly used in the manufacture of catheters with a melting point of around 200 °C. They are popular in the manufacture of catheters for dialysis as well as urinary catheters, due to their flexibility, good blood compatibility, and strength. In fact, the incidence of catheter-related bloodstream infection is lower for PU catheters than those made of PVC or PU and so it is the current material of choice in manufacturing of catheters, particularly due to its potential to form sustained release polymers through HME, which has been proven in medical tubing [78]. One study in particular outlines the use of TPU in HME and its potential for use in FDM printing [79]. To conclude, TPU filament is suitable for 3D printing and has potential for creating customised and repeatable products.

4.5. Silicone

Silicone, is an inert polymer with good thermal stability, moisture resistance, and flexibility [80]. An alternative name for this type of polymer is poly(siloxane) (Figure 6). Silicone has a glass transition at around −127 °C and a melting point at around −43 °C [81]. It is resistant to temperature from −55 to around 300 °C. Silicone has been extensively used for the production of catheters; for example, on the production of antimicrobial graphene nanoplatelet coatings for silicone catheters, and on the surface modification of silicone with colloidal polysaccharide formulations for the development of antimicrobial urethral catheters. Silicone has been proven to have effective use in FDM printing [82,83]. There was a study done to show the effective use of moisture-cured silicone in extrusion based AM [82]. It also has minimal leaching of plasticisers from its matrix.

4.6. Latex

Latex is used in the manufacture of catheters, as it is a soft flexible material with a melting point of around 180 °C. Latex has a high stretch ratio, is impermeable to water, and is a resilient material [84]. This is the original material used in the manufacture of Foley urinary catheters. An issue with latex catheters is cytotoxicity, due to elute from the rubber. Therefore, latex is not as commonly used today as PU or silicone catheters. Some latex catheters may be covered with a layer of silicone in order to minimise cytotoxicity. However, as latex is a rubber, it is not ideal for the processes of HME and FDM. Once latex becomes molten and sets, it cannot be melted again. Therefore, latex is not considered as a thermoplastic elastomer.

5. Regulatory Considerations on 3D-Printed Medical Devices

Since the FDA first approved a 3D-printed drug (Spritam®, Aprecia Pharmaceuticals, Blue Ash, OH, USA) in 2015, there has been a growing interest in 3D printing of pharmaceutical products and devices. This has also provided a major breakthrough in the regulation of 3D-printed pharmaceuticals [85]. 3D printing offers the potential to produce multiple devices daily using one process with flexibility within the device design process. However, this also opens up the opportunity for variability to occur [86]. Therefore, the FDA have released guidance documentation on AM of medical devices. The guidance focuses around design and manufacturing considerations and device testing considerations for 3D-printed medical devices. The guidance covers areas such as file format conversions, validating and automating software processes, material controls, device testing considerations, and material characterisation. As the potential to create patient-matched devices is a major advantage of AM, the FDA states that when imaging a patient, a risk-based approach should be used to assess scenarios in which a worst-case match to the patient would be produced. Quality must be maintained for all devices by performing process validation for all devices and components built in a single build cycle, between build cycles, and between machines. For all validated processes, there should be clear documentation on the data and monitoring and control methods. Revalidation must be performed when there have been any changes made to the manufacturing process to assess any risks the changes may bring. Throughout the documentation, there are also references to current guidance that already exists for medical devices; therefore, the guidance provided on AM should be supplemented with existing guidance on a particular medical device [87].

Author Contributions: Conceptualization, E.M., J.D.-R., E.L. and D.A.L.; Investigation, E.M., J.D.-R. and E.L.; Writing—original draft preparation, E.M., J.D.-R. and E.L.; Writing—review and editing, J.D.-R., E.L. and D.A.L.; Supervision, E.L. and D.A.L.

Funding: This research received no external funding.

Conflicts of Interest: The authors declare no conflict of interest.

References

1. Kidney Care UK. *Patient Reported Experience of Kidney Care in England and Wales 2017*; Kidney Care UK: Alton, UK, 2017.
2. Couser, W.G.; Remuzzi, G.; Mendis, S.; Tonelli, M. The contribution of chronic kidney disease to the global burden of major noncommunicable diseases. *Kidney Int.* **2011**, *80*, 1258–1270. [CrossRef] [PubMed]
3. Kidney Care UK. Facts and Stats. Available online: https://www.kidneycareuk.org/news-and-campaigns/facts-and-stats/ (accessed on 7 March 2019).
4. Kerr, M. *Chronic Kidney Disease in England: The Human and Financial Cost*; Insight Health Economics Ltd.: Twickenham, UK, 2012; p. 10.
5. Lewis, R. *Understanding Chronic Kidney Disease: A Guide for the Non-Specialist*; M&K Update Ltd.: Keswick, UK, 2011.
6. Foreman, K.J.; Marquez, N.; Dolgert, A.; Fukutaki, K.; Fullman, N.; McGaughey, M.; Pletcher, M.A.; Smith, A.E.; Tang, K.; Yuan, C.W.; et al. Forecasting life expectancy, years of life lost, and all-cause and cause-specific mortality for 250 causes of death: Reference and alternative scenarios for 2016–40 for 195 countries and territories. *Lancet* **2018**, *392*, 2052–2090. [CrossRef]
7. Dialysis—NHS. Available online: https://www.nhs.uk/conditions/dialysis/ (accessed on 7 March 2019).
8. Miller, D.L.; O'Grady, N.P. Guidelines for the Prevention of Intravascular Catheter-related Infections: Recommendations Relevant to Interventional Radiology for Venous Catheter Placement and Maintenance. *J. Vasc. Interv. Radiol.* **2012**, *23*, 997–1007. [CrossRef] [PubMed]
9. Kramer, R.D.; Rogers, M.A.M.; Conte, M.; Mann, J.; Saint, S.; Chopra, V. Are antimicrobial peripherally inserted central catheters associated with reduction in central line—associated bloodstream infection? A systematic review and meta-analysis. *AJIC Am. J. Infect. Control* **2017**, *45*, 108–114. [CrossRef] [PubMed]
10. Mendu, M.L.; May, M.F.; Kaze, A.D.; Graham, D.A.; Cui, S.; Chen, M.E.; Shin, N.; Aizer, A.A.; Waikar, S.S. Non-Tunneled versus tunneled dialysis catheters for acute kidney injury requiring renal replacement therapy: A prospective cohort study. *BMC Nephrol.* **2017**, *18*, 1–7. [CrossRef]
11. van Oevelen, M.; Abrahams, A.C.; Weijmer, M.C.; Nagtegaal, T.; Dekker, F.W.; Rotmans, J.I.; Meijvis, S.C.A.; Bijlsma, J.A.; van der Bogt, K.E.A.; van de Brug, A.; et al. Precurved non-tunnelled catheters for haemodialysis are comparable in terms of infections and malfunction as compared to tunnelled catheters: A retrospective cohort study. *J. Vasc. Access* **2019**, *20*, 307–312. [CrossRef] [PubMed]
12. Sutherland, D.W.; Zhang, X.; Charest, J.L. Water infused surface protection as an active mechanism for fibrin sheath prevention in central venous catheters. *Artif. Organs* **2017**, *41*, E155–E165. [CrossRef]
13. Ripa, M.; Morata, L.; Rodríguez-Núñez, O.; Cardozo, C.; Puerta-Alcalde, P.; Hernández-Meneses, M.; Ambrosioni, J.; Linares, L.; Bodro, M.; Valcárcel, A.; et al. Short-term peripheral venous catheter-related bloodstream infections. Evidence for increasing prevalence of gram-negative microorganisms from a 25-year prospective observational study. *Antimicrob. Agents Chemother.* **2018**, *62*, e00892-18. [CrossRef]
14. Miller, L.M.; Clark, E.; Dipchand, C.; Hiremath, S.; Kappel, J.; Kiaii, M.; Lok, C.; Luscombe, R.; Moist, l. Hemodialysis tunneled catheter noninfectious complications. *Can. J. Kidney Health Dis.* **2016**, *3*, 2054358116669130. [CrossRef]
15. Wiese, L.; Mejer, N.; Schønheyder, H.C.; Westh, H. A nationwide study of comorbidity and risk of reinfection after Staphylococcus aureus bacteraemia. *J. Infect.* **2013**, *67*, 199–205. [CrossRef]

16. Brass, P.; Hellmich, M.; Kolodziej, L.; Schick, G.; Af, S. Ultrasound guidance versus anatomical landmarks for internal jugular vein catheterization. *Cochrane Database Syst. Rev.* **2015**. [CrossRef] [PubMed]

17. Wang, H.; Tong, H.; Liu, H.; Wang, Y.; Wang, R.; Gao, H.; Yu, P.; Lv, Y.; Chen, S.; Wang, G.; et al. Effectiveness of antimicrobial-coated central venous catheters for preventing catheter-related blood-stream infections with the implementation of bundles: A systematic review and network meta-analysis. *Ann. Intensive Care* **2018**, *8*, 71. [CrossRef] [PubMed]

18. Zanini, S.; Polissi, A.; Maccagni, E.A.; Dell'Orto, E.C.; Liberatore, C.; Riccardi, C. Development of antibacterial quaternary ammonium silane coatings on polyurethane catheters. *J. Colloid Interface Sci.* **2015**, *451*, 78–84. [CrossRef] [PubMed]

19. Schneider, A.; Baldwin, I.; Souweine, B. What's new: Prevention of acute dialysis catheter-related infection. *Intensive Care Med.* **2018**, *44*, 356–358. [CrossRef] [PubMed]

20. Keum, H.; Kim, J.Y.; Yu, B.; Yu, S.J.; Kim, J.; Jeon, H.; Lee, D.Y.; Im, S.G.; Jon, S. Prevention of bacterial colonization on catheters by a one-step coating process involving an antibiofouling polymer in water. *ACS Appl. Mater. Interfaces* **2017**, *9*, 19736–19745. [CrossRef]

21. Lok, C.E.; Mokrzycki, M.H. Prevention and management of catheter-related infection in hemodialysis patients. *Kidney Int.* **2011**, *79*, 587–598. [CrossRef]

22. Mogi, N.; Nakagawa, M.; Matsumae, H.; Hattori, A.; Shimohira, M.; Shibamoto, Y. Fibrin sheath of a peripherally inserted central catheter undepicted with gray-scale (real-time B-mode) ultrasonography: A case report. *Radiol. Case Rep.* **2018**, *13*, 537–541. [CrossRef] [PubMed]

23. Valliant, A.M.; Chaudhry, M.K.; Yevzlin, A.; Astor, B.; Chan, M.R. Tunneled dialysis catheter exchange with fibrin sheath disruption is not associated with increased rate of bacteremia. *J. Vasc. Access* **2015**, *16*, 52–56. [CrossRef]

24. Wang, L.H.; Wei, F.; Jia, L.; Lu, Z.; Wang, B.; Dong, H.Y.; Yu, H.B.; Sun, G.J.; Yang, J.; Li, B.; et al. Fibrin sheath formation and intimal thickening after catheter placement in dog model: Role of hemodynamic wall shear stress. *J. Vasc. Access* **2015**, *16*, 275–284. [CrossRef]

25. Ahmed, R.; Chapman, S.A.; Tantrige, P.; Hussain, A.; Johnston, E.W.; Fang, C.; Ammar, T.; Huang, D.Y.; Wilkins, C.J.; Garzillo, G.; et al. TuLIP (Tunnelled Line Intraluminal Plasty): An alternative technique for salvaging haemodialysis catheter patency in fibrin sheath formation. *Cardiovasc. Interv. Radiol.* **2019**, *42*, 770–774. [CrossRef]

26. May, R.M.; Brennan, A.B.; Fraser, J.C.; Drinker, M.C.; Mann, E.E.; Reddy, S.T.; Magin, C.M.; Siedlecki, C.A. An engineered micropattern to reduce bacterial colonization, platelet adhesion and fibrin sheath formation for improved biocompatibility of central venous catheters. *Clin. Transl. Med.* **2015**, *4*, 1–8. [CrossRef]

27. UV-CURE SILICONES ENABLE NEW MEDICAL DEVICE DESIGN CONCEPTS AND INCREASE CUSTOMER VALUE. Available online: https://www.freudenbergmedical.com/ecomaXL/files/UV-Cure_Silicones.pdf?download=1 (accessed on 14 August 2019).

28. Weisenberg, S.H.; Termaath, S.C.; Seaver, C.E.; Killeffer, J.A. Ventricular catheter development: Past, present, and future. *J. Neurosurg.* **2016**, *125*, 1504–1512. [CrossRef]

29. Yang, S.; Lee, Y.J.; Lin, F.; Yang, J.; Chen, K. Chitosan/Poly (vinyl alcohol) blending hydrogel coating improves the surface characteristics of segmented polyurethane urethral catheters. *J. Biomed. Mater. Res. Part B* **2007**, *83*, 304–313. [CrossRef]

30. Weisman, J.A.; Ballard, D.H.; Jammalamadaka, U.; Tappa, K.; Sumerel, J.; D'Agostino, H.B.; Mills, D.K.; Woodard, P.K. 3D printed antibiotic and chemotherapeutic eluting catheters for potential use in interventional radiology: In vitro proof of concept study. *Acad. Radiol.* **2019**, *26*, 270–274. [CrossRef]

31. Balne, P.K.; Harini, S.; Dhand, C.; Dwivedi, N.; Chalasani, M.L.S.; Verma, N.K.; Barathi, V.A.; Beuerman, R.; Agrawal, R.; Lakshminarayanan, R. Surface characteristics and antimicrobial properties of modified catheter surfaces by polypyrogallol and metal ions. *Mater. Sci. Eng. C* **2018**, *90*, 673–684. [CrossRef]

32. Falk, A. The Role of Surface Coatings on Central Venous and Hemodialysis Catheters. *Endovasc. Today* **2008**, 28–30.

33. Ps, S.; Shah, N. Heparin-bonded catheters for prolonging the patency of central venous catheters in children. *Cochrane Database Syst. Rev.* **2014**, 3–5. [CrossRef]

34. Nicolas, D.R.M. *Heparin Induced Thrombocytopenia (HIT)*; StatPearls: Treasure Island, FL, USA, 2018.

35. Kasirajan, K. Outcomes after heparin-Induced thrombocytopenia in patients with propaten vascular grafts. *Ann. Vasc. Surg.* **2012**, *26*, 802–808. [CrossRef]

36. Franci, G.; Falanga, A.; Galdiero, S.; Palomba, L.; Rai, M.; Morelli, G.; Galdiero, M. Silver nanoparticles as potential antibacterial agents. *Molecules* **2015**, *20*, 8856–8874. [CrossRef]
37. Reidy, B.; Haase, A.; Luch, A.; Dawson, A.K.; Lynch, I. Mechanisms of silver nanoparticle release, transformation and toxicity: A critical review of current knowledge and recommendations for future studies and applications. *Materials* **2013**, *6*, 2295–2350. [CrossRef]
38. Wu, K.; Yang, Y.; Zhang, Y.; Deng, J.; Lin, C. A ntimicrobial activity and cytocompatibility of silver nanoparticles coated catheters via a biomimetic surface functionalization strategy. *Int. J. Nanomed.* **2015**, *10*, 7241–7252. [CrossRef]
39. Kuehl, R.; Brunetto, P.S.; Woischnig, A.K.; Varisco, M.; Rajacic, Z.; Vosbeck, J.; Terracciano, L.; Fromm, K.M.; Khanna, N. Preventing implant-associated infections by silver coating. *Antimicrob. Agents Chemother.* **2016**, *60*, 2467–2475. [CrossRef]
40. Freitas, P.D. Incorporation of Silver Nanoparticles by the Irradiation Process in Central Venous Catheter (CVC) of Polyurethane Coated with Titanium Oxide for Antimicrobial Activity. Master's Thesis, Instituto de Pesquisas Energéticas e Nucleares, Sao Paulo, Brazil, July 2018.
41. Chen, Y.M.; Dai, A.P.; Shi, Y.; Liu, Z.J.; Gong, M.F.; Yin, X.B. Effectiveness of silver-impregnated central. venous catheters for preventing catheter-related blood stream infections: A meta-analysis. *Int. J. Infect. Dis.* **2014**, *29*, e279–e286. [CrossRef]
42. Iqtedar, M.; Aslam, M.; Akhyar, M.; Shehzaad, A.; Abdullah, R.; Kaleem, A. Extracellular biosynthesis, characterization, optimization of silver nanoparticles (AgNPs) using Bacillus mojavensis BTCB15 and its antimicrobial activity against multidrug resistant pathogens. *Prep. Biochem. Biotechnol.* **2019**, *49*, 136–142. [CrossRef]
43. Roe, D.; Karandikar, B.; Bonn-Savage, N.; Gibbins, B.; Roullet, J. baptiste Antimicrobial surface functionalization of plastic catheters by silver nanoparticles. *J. Antimicrob. Chemother.* **2008**, *61*, 869 876. [CrossRef]
44. Kubo, A.L.; Capjak, I.; Vrček, I.V.; Bondarenko, O.M.; Kurvet, I.; Vija, H.; Ivask, A.; Kasemets, K.; Kahru, A. Antimicrobial potency of differently coated 10 and 50 nm silver nanoparticles against clinically relevant bacteria Escherichia coli and Staphylococcus aureus. *Colloids Surf. B Biointerfaces* **2018**, *170*, 401–410. [CrossRef]
45. Kim, S.H.; Lee, H.S.; Ryu, D.S.; Choi, S.J.; Lee, D.S. Antibacterial activity of silver-nanoparticles against Staphylococcus aureus and Escherichia coli. *Korean J. Microbiol. Biotechnol.* **2011**, *39*, 77–85.
46. Chopra, I. The increasing use of silver-based products as antimicrobial agents: A useful development or a cause for concern? *J. Antimicrob. Chemother.* **2007**, *59*, 587–590. [CrossRef]
47. Percival, S.L.; Bowler, P.G.; Russell, D. Bacterial resistance to silver in wound care. *J. Hosp. Infect.* **2005**, *60*, 1–7. [CrossRef]
48. Silver, S. Bacterial silver resistance: Molecular biology and uses and misuses of silver compounds. *FEMS Microbiol. Rev.* **2003**, *27*, 341–353. [CrossRef]
49. Korani, M.; Ghazizadeh, E.; Korani, S.; Hami, Z.; Mohammadi-Bardbori, A. Effects of silver nanoparticles on human health. *Eur. J. Nanomed.* **2015**, *7*, 51–62. [CrossRef]
50. Stenger, M.; Klein, K.; Grønnemose, R.B.; Klitgaard, J.K.; Kolmos, H.J.; Lindholt, J.S.; Alm, M.; Thomsen, P.; Andersen, T.E. Co-release of dicloxacillin and thioridazine from catheter material containing an interpenetrating polymer network for inhibiting device-associated Staphylococcus aureus infection. *J. Control. Release* **2016**, *241*, 125–134. [CrossRef]
51. Zanwar, S.; Jain, P.; Gokarn, A.; Kumar, S.; Punatar, S.; Khurana, S.; Bonda, A.; Pruthy, R.; Bhat, V.; Qureshi, S.; et al. Antibiotic lock therapy for salvage of tunneled central venous catheters with catheter colonization and catheter-related bloodstream infection. *Transpl. Infect. Dis.* **2019**, *21*, e13017. [CrossRef]
52. Melocchi, A.; Parietti, F.; Maroni, A.; Foppoli, A.; Gazzaniga, A.; Zema, L. Hot-melt extruded filaments based on pharmaceutical grade polymers for 3D printing by fused deposition modeling. *Int. J. Pharm.* **2016**, *509*, 255–263. [CrossRef]
53. Weisman, J.A.; Nicholson, J.C.; Tappa, K.; Jammalamadaka, U.; Wilson, C.G.; Mills, D.K. Antibiotic and chemotherapeutic enhanced three-dimensional printer filaments and constructs for biomedical applications. *Int. J. Nanomed.* **2015**, *10*, 357–370. [CrossRef]
54. Sousa, C.; Teixeira, P.; Oliveira, R. Influence of surface properties on the adhesion of staphylococcus epidermidis to acrylic and silicone. *Int. J. Biomater.* **2009**, *2009*, 1–9. [CrossRef]

55. Lalehpour, A.; Janeteas, C.; Barari, A. Surface roughness of FDM parts after post-processing with acetone vapor bath smoothing process. *Int. J. Adv. Manuf. Technol.* **2018**, *95*, 1505–1520. [CrossRef]

56. Pérez, M.; Medina-Sánchez, G.; García-Collado, A.; Gupta, M.; Carou, D. Surface quality enhancement of fused deposition modeling (FDM) printed samples based on the selection of critical printing parameters. *Materials* **2018**, *11*, 1382. [CrossRef]

57. Keating, A.V.; Soto, J.; Tuleu, C.; Forbes, C.; Zhao, M.; Craig, D.Q.M. Solid state characterisation and taste masking efficiency evaluation of polymer based extrudates of isoniazid for paediatric administration. *Int. J. Pharm.* **2018**, *536*, 536–546. [CrossRef]

58. Verstraete, G.; Van Renterghem, J.; Van Bockstal, P.-J.; Kasmi, S.; De Geest, B.; De Beer, T.; Remon, J.P.; Vervaet, C. Hydrophilic thermoplastic polyurethanes for the manufacturing of highly dosed oral sustained release matrices via hot melt extrusion and injection molding. *Int. J. Pharm.* **2016**, *506*, 214–221. [CrossRef]

59. Muwaffak, Z.; Goyanes, A.; Clark, V.; Basit, A.W.; Hilton, S.T.; Gaisford, S. Patient-specific 3D scanned and 3D printed antimicrobial polycaprolactone wound dressings. *Int. J. Pharm.* **2017**, *527*, 161–170. [CrossRef]

60. Tan, D.K.; Maniruzzaman, M.; Nokhodchi, A. Advanced pharmaceutical applications of hot-melt extrusion coupled with fused deposition modelling (FDM) 3D printing for personalised drug delivery. *Pharmaceutics* **2018**, *10*, 203. [CrossRef]

61. Alhijjaj, M.; Belton, P.; Qi, S. An investigation into the use of polymer blends to improve the printability of and regulate drug release from pharmaceutical solid dispersions prepared via fused deposition modeling (FDM) 3D printing. *Eur. J. Pharm. Biopharm.* **2016**, *108*, 111–125. [CrossRef]

62. Tappa, K.; Jammalamadaka, U. Novel biomaterials used in medical 3D printing techniques. *J. Funct. Biomater.* **2018**, *9*, 17. [CrossRef]

63. Stewart, A.S.; Domínguez-Robles, J.; Donnelly, F.R.; Larrañeta, E. Implantable polymeric drug delivery devices: Classification, manufacture, materials, and clinical applications. *Polymers* **2018**, *10*, 1379. [CrossRef]

64. Domínguez-Robles, J.; Martin, K.N.; Fong, L.M.; Stewart, A.S.; Irwin, J.N.; Rial-Hermida, I.M.; Donnelly, F.R.; Larrañeta, E. Antioxidant PLA composites containing lignin for 3d printing applications: A potential material for healthcare applications. *Pharmaceutics* **2019**, *11*, 165. [CrossRef]

65. Pawar, R.P.; Tekale, S.U.; Shisodia, S.U.; Totre, J.T.; Domb, A.J. Biomedical applications of Poly (Lactic Acid). *Rec. Pat. Regen. Med.* **2014**, *4*, 40–51. [CrossRef]

66. Jacobi, C.; Friedrich, T.; Lüdtke-Buzug, K. Synthesis and characterisation of superparamagnetic polylactic acid based polymers. *Int. J. Magn. Part. Imaging* **2017**, *3*, 1710001.

67. Masutani, K.; Kimura, Y. PLA Synthesis and Polymerization. In *Poly(Lactic Acid) Science and Technology: Processing, Properties, Additives and Applications*; Jiménez, A., Peltzer, M., Ruseckaite, R., Eds.; Royal Society of Chemistry: London, UK, 2014; ISBN 9781782624806.

68. Farah, S.; Anderson, D.G.; Langer, R. Physical and mechanical properties of PLA, and their functions in widespread applications—A comprehensive review. *Adv. Drug Deliv. Rev.* **2016**, *107*, 367–392. [CrossRef]

69. Romanov, V.; Samuel, R.; Chaharlang, M.; Jafek, A.R.; Frost, A.; Gale, B.K. FDM 3D printing of high-pressure, heat-resistant, transparent microfluidic devices. *Anal. Chem.* **2018**, *90*, 10450–10456. [CrossRef]

70. Guerra, A.J.; Cano, P.; Rabionet, M.; Puig, T.; Ciurana, J. 3D-printed PCL/PLA composite stents: Towards a new solution to cardiovascular problems. *Materials* **2018**, *11*, 1679. [CrossRef]

71. Li, D.; Guo, G.; Fan, R.; Liang, J.; Deng, X.; Luo, F.; Qian, Z. PLA/F68/Dexamethasone implants prepared by hot-melt extrusion for controlled release of anti-inflammatory drug to implantable medical devices: I. Preparation, characterization and hydrolytic degradation study. *Int. J. Pharm.* **2013**, *441*, 365–372. [CrossRef]

72. Li, D.; Guo, G.; Deng, X.; Fan, R.; Guo, Q.; Fan, M.; Liang, J.; Luo, F.; Qian, Z. PLA/PEG-PPG-PEG/Dexamethasone implant prepared by hot-melt extrusion for controlled release of immunosuppressive drug to implantable medical devices, part 2: In vivo evaluation. *Drug Deliv.* **2013**, *20*, 134–142. [CrossRef]

73. Visscher, L.E.; Dang, H.P.; Knackstedt, M.A.; Hutmacher, D.W.; Tran, P.A. 3D printed Polycaprolactone scaffolds with dual macro-microporosity for applications in local delivery of antibiotics. *Mater. Sci. Eng. C* **2018**, *87*, 78–89. [CrossRef]

74. Fu, J.; Yu, X.; Jin, Y. 3D printing of vaginal rings with personalized shapes for controlled release of progesterone. *Int. J. Pharm.* **2018**, *539*, 75–82. [CrossRef]

75. Johansson, K.; Greis, G.; Johansson, B.; Grundtmann, A.; Pahlby, Y.; Törn, S.; Axelberg, H.; Carlsson, P. Evaluation of a new PVC-free catheter material for intermittent catheterization: A prospective, randomized, crossover study. *Scand. J. Urol.* **2013**, *47*, 33–37. [CrossRef]

76. Sharma, R.; Singh, R.; Penna, R.; Fraternali, F. Investigations for mechanical properties of Hap, PVC and PP based 3D porous structures obtained through biocompatible FDM filaments. *Compos. Part B Eng.* **2018**, *132*, 237–243. [CrossRef]

77. Rusu, M.; Ursu, M.; Rusu, D. Poly(vinyl chloride) and poly(e-caprolactone) blends for medical use. *J. Thermoplast. Compos. Mater.* **2006**, *19*, 173–190. [CrossRef]

78. Simons, F.J.; Wagner, K.G. Modeling, design and manufacture of innovative floating gastroretentive drug delivery systems based on hot-melt extruded tubes. *Eur. J. Pharm. Biopharm.* **2019**, *137*, 196–208. [CrossRef]

79. Haryńska, A.; Gubanska, I.; Kucinska-Lipka, J.; Janik, H. Fabrication and characterization of flexible medical-grade tpu filament for fused deposition modeling 3dp technology. *Polymers* **2018**, *10*, 1304. [CrossRef]

80. Gupta, S.; Ramamurthy, P.C.; Madras, G. Synthesis and characterization of silicone polymer/functionalized mesostructured silica composites. *Polym. Chem.* **2011**, *2*, 2643–2650. [CrossRef]

81. Thermal Analysis of Silicone Rubber. Available online: https://www.hitachi-hightech.com/file/global/pdf/products/science/appli/ana/thermal/application_TA_018e.pdf (accessed on 14 August 2018).

82. Plott, J.; Shih, A. The extrusion-based additive manufacturing of moisture-cured silicone elastomer with minimal void for pneumatic actuators. *Addit. Manuf.* **2017**, *17*, 1–14. [CrossRef]

83. Hamidi, A.; Jain, S.; Tadesse, Y. 3D printing PLA and silicone elastomer structures with sugar solution support material. In Proceedings of the SPIE, Portland, OR, USA, 17 April 2017; Volume 10163.

84. Feneley, R.C.L.; Hopley, I.B.; Wells, P.N.T. Urinary catheters: History, current status, adverse events and research agenda. *J. Med. Eng. Technol.* **2015**, *39*, 459–470. [CrossRef]

85. Khatri, P.; Shah, M.K.; Vora, N. Formulation strategies for solid oral dosage form using 3D printing technology: A mini-review. *J. Drug Deliv. Sci. Technol.* **2018**, *46*, 148–155. [CrossRef]

86. Adamo, J.E.; Grayson, W.L.; Hatcher, H.; Swanton, J.; Thomas, A.; Hollister, S.; Steele, S.J. Regulatory interfaces surrounding the growing field of additive manufacturing of medical devices and biologic products. *J. Clin. Transl. Sci.* **2018**, *2*, 301–304. [CrossRef]

87. U.S. Food and Drug Administration. *Technical Considerations for Additive Manufactured Medical Devices. Guidance for Industry and Food and Drug Administration Staff*; U.S. Food and Drug Administration: Silver Spring, MD, USA, 2017.

Review

Edible Films and Coatings for Fresh Fish Packaging: Focus on Quality Changes and Shelf-life Extension

Maria-Ioana Socaciu [1], Cristina Anamaria Semeniuc [2],*and Dan Cristian Vodnar [1],*

[1]   Department of Food Science, University of Agricultural Sciences and Veterinary Medicine Cluj-Napoca, 3-5 Mănăştur St., 400372 Cluj-Napoca, Romania; maria-ioana.socaciu@usamvcluj.ro

[2]   Department of Food Engineering, University of Agricultural Sciences and Veterinary Medicine Cluj-Napoca, 3-5 Mănăştur St., 400372 Cluj-Napoca, Romania

*   Correspondence: cristina.semeniuc@usamvcluj.ro (C.A.S.); dan.vodnar@usamvcluj.ro (D.C.V.); Tel.: +40-264-596-384 (C.A.S. & D.C.V.)

Received: 16 August 2018; Accepted: 13 October 2018; Published: 16 October 2018

Abstract: Fresh fish is extensively consumed and is one of the most-traded food commodities in the world. Conventional preservation technologies include vacuum and modified atmosphere packaging, but they are costly since requires capital investment. In the last decade, research has been directed towards the development of antimicrobial packaging systems, as an economical alternative to these. This paper outlines antimicrobial films and coatings applied so far on fresh fish, their efficacy against targeted microorganism/group and effects on chemical quality of the product. Findings show that edible films/coatings incorporated with different active agents applied to fresh fish are able to inhibit the microbial growth and decrease the rate of fish nutrients degradation, thus preventing the formation of chemical metabolites; a shelf-life extension of 6 to 13 days was obtained for fish fillets, depending on the species on which the active packaging materials were applied. The manufacturing use of these formulations could lead to a significant reduction in fish waste, consequently, a diminution of economic losses for fish traders and retailers. Therefore, their industrial production and commercialization could be an exploitable sector by the packaging industry.

Keywords: edible films; edible coatings; antimicrobial agents; fresh fish; spoilage; shelf-life

---

1. Introduction

Fish is one of the most-traded food commodities worldwide [1]. Capture fisheries and aquaculture provide valuable economic and social benefits to those who work in these industries [2]. However, post-harvest handling, processing, and storage of fish lead to food losses and waste [3]. Post-harvest losses occur at all stages in the fish supply chain from capture to consumer [4]. The losses can be physical, economical, or nutritional and are caused by spoilage or poor processing [5]. Spoilage is the process in which fish deteriorates to the point that becomes unacceptable for human consumption (with altered taste, smell, appearance, or texture) [6]. Globally, fish losses that are caused by spoilage account for around 10% (10 to 12 million tons per year) of the total production from capture fisheries and aquaculture [7].

Fresh fish is a highly perishable product due to its high water activity, nutrient availability, nearly neutral-pH (factors that influence microbial growth) and the presence of autolytic enzymes; hence, it is susceptible to post-harvest losses [8,9]. Under normal refrigerated storage conditions, its shelf-life is limited by the development of enzymatic (caused by endogenous or microbial enzymes) and chemical reactions [10]. The main initial causative factor for fish spoilage is microbial growth and invasion, followed by the autolytic enzymes and then by chemical reactions, such as oxidation or hydrolysis [11,12].

Post-harvest losses of fresh fish due to microbial spoilage are a matter of great importance to the fishing industry [13]. So, specific requirements and preservation techniques are needed to minimize the activity of spoilage bacteria. Fresh fish products are presently stored on ice or under refrigeration during their distribution and marketing. In these conditions, their shelf-life is limited to 5–10 days (depending on species, harvest location, and season) and they can result in enormous economic losses to fish traders and retailers [14,15]. Therefore, the fish-process industry is actively seeking alternative methods of shelf-life preservation and marketability of fresh fish [16].

Packaging plays a critical role in the fish supply chain and is part of the solution to tackle food waste [17,18]. Vacuum packaging (VP) and modified atmosphere packaging (MAP) are very commonly used as a supplement to ice or refrigeration to inhibit the normal spoilage flora and extend the shelf-life of fresh fish products [14,19,20]. MAP technology has, however, some disadvantages, such as added costs for packaging equipment, gases, and packaging materials; it also requires special training for food operators [21].

Packaging innovation and new technologies is a necessity for the fishing industry. In recent years, a variety of active packaging systems have been developed to prolong storage life and enhance the safety of fish products. These have a variety of advantages such as biodegradability, edibility, biocompatibility, and aesthetic appearance, respectively, barrier properties against oxygen and physical stress [22]. The purpose of this paper is to provide an overview of published research about edible films and coatings applied to fresh fish. The antimicrobial films and coatings that are used for fish packaging and their effects on chemical quality of fresh fish are reviewed and discussed (Figure 1).

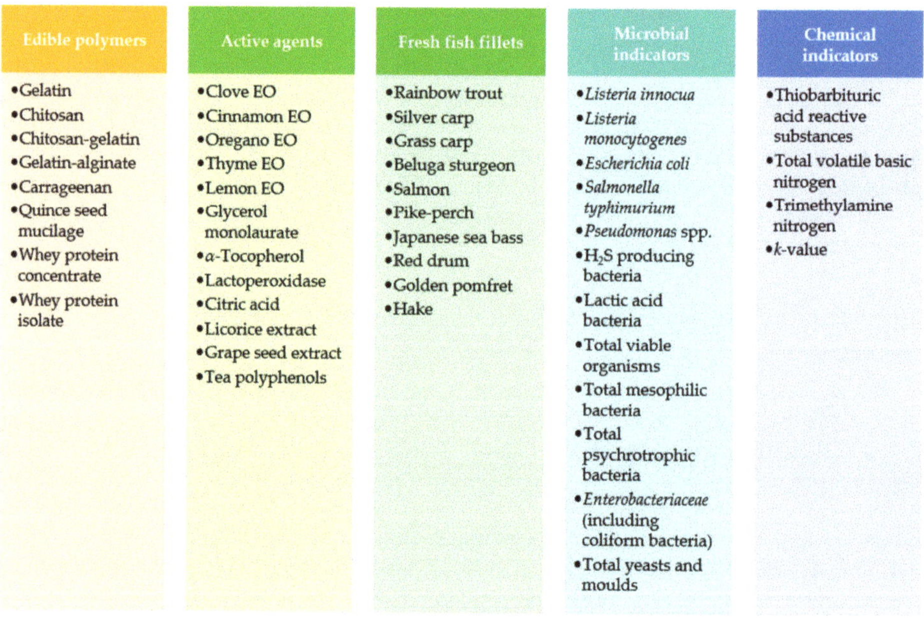

Edible polymers	Active agents	Fresh fish fillets	Microbial indicators	Chemical indicators
• Gelatin	• Clove EO	• Rainbow trout	• *Listeria innocua*	• Thiobarbituric acid reactive substances
• Chitosan	• Cinnamon EO	• Silver carp	• *Listeria monocytogenes*	• Total volatile basic nitrogen
• Chitosan-gelatin	• Oregano EO	• Grass carp	• *Escherichia coli*	• Trimethylamine nitrogen
• Gelatin-alginate	• Thyme EO	• Beluga sturgeon	• *Salmonella typhimurium*	• *k*-value
• Carrageenan	• Lemon EO	• Salmon	• *Pseudomonas* spp.	
• Quince seed mucilage	• Glycerol monolaurate	• Pike-perch	• H_2S producing bacteria	
• Whey protein concentrate	• α-Tocopherol	• Japanese sea bass	• Lactic acid bacteria	
• Whey protein isolate	• Lactoperoxidase	• Red drum	• Total viable organisms	
	• Citric acid	• Golden pomfret	• Total mesophilic bacteria	
	• Licorice extract	• Hake	• Total psychrotrophic bacteria	
	• Grape seed extract		• *Enterobacteriaceae* (including coliform bacteria)	
	• Tea polyphenols		• Total yeasts and moulds	

Figure 1. Antimicrobial films and coatings used to extend the shelf-life of fresh fish fillets.

2. Microbiological Issues

Fresh fish spoils due to the action of a group of microorganisms, the so-called specific spoilage organisms (SSOs). These organisms have the ability to dominate the fish flora and produce metabolites that directly affect the sensory properties of the product resulting in its rejection by consumers [23]. During storage, the microflora changes owing to different capacities of the microorganisms to tolerate the preservation conditions [24]. Under aerobic iced storage, the flora of fish is composed almost exclusively of *Pseudomonas* spp. and *Shewanella putrefaciens* (SSOs) regardless of whether it was caught or harvested in temperate or sub-tropical and tropical waters. At ambient temperature (25 °C), microflora is dominated by mesophilic *Vibrionaceae*, and, particularly if the fish is caught in polluted waters, by mesophilic *Enterobacteriaceae* [25].

Microbial spoilage is due to the proliferation of microorganisms after the death of fish as a result of the immune system collapsing, followed by the microbial invasion of the fish body through the skin [12]. Fish have a unique osmoregulatory mechanism to avoid dehydration in marine environments and waterlogging of tissue in freshwater; it contains osmoregulatory compounds, like trimethylamine oxide (TMAO) and urea [26]. Microbial enzymes that are present in fish can break down TMAO to trimethylamine (TMA) and urea to ammonia, volatile organic compounds associated with microbial spoilage [12]. Many other volatile compounds can be formed by microbial enzymatic degradation of other substrates, such as hydrogen sulphide (from cysteine), methanethiol and methyl sulphide (from methionine), histamine (from histidine), acetate, carbon dioxide and water (from carbohydrates and lactate), hypoxanthine (from inosine and inosine-5′-monophosphate), esters, ketones, aldehydes (from amino acids, like glycine, serine, and leucine), as well as ammonia (from amino acids and urea) [12,26]. These molecules are responsible for sweet, fruity, ammonia-like, putrid, and sulphuric off-flavours in spoiled fish [27].

3. Antimicrobial Films and Coatings Applied on Fresh Fish

This chapter provides an overview of previous research on the antimicrobial packaging of fresh fish. Table 1 lists active edible films and coatings applied to fresh fish fillets (of rainbow trout, silver carp, grass carp, beluga sturgeon, salmon, pike-perch, Japanese sea bass, red drum, golden pomfret, and hake) to extend its shelf-life. These films and coatings were produced from edible polymers like gelatin, chitosan, chitosan-gelatin, gelatin-alginate, carrageenan, quince seed mucilage, whey protein concentrate, and whey protein isolate incorporated with various active agents (essential oils (EOs) of clove, cinnamon, oregano, thyme, and lemon, glycerol monolaurate, α-tocopherol, lactoperoxidase, citric acid, licorice extract, grape seed extract, and tea polyphenols). Their antimicrobial efficacy was investigated in situ against spoilage and pathogenic microorganisms. Different levels of effectiveness were noticed, depending on the active agent used, its concentration, storage temperature, atmosphere composition (normal or modified), and targeted microorganism/group.

3.1. Efficacy against Tested Microorganism/Group at the End of Monitoring Time

3.1.1. Efficacy against Spoilage Microorganisms

Several authors have investigated the potential of edible films/coatings in extending the shelf-life of fresh fish fillets by retarding the growth of spoilage bacteria. Jouki et al. (2014) [28] have tested the efficacy of films based on 1% quince seed mucilage incorporated with different concentrations of oregano and thyme EOs (1%, 1.5%, and 2%) against *Pseudomonas* spp., H₂S producing bacteria, and lactic acid bacteria in rainbow trout fillets; Kazemi & Rezaei (2015) [29] of films based on 3% gelatin and 1.5% alginate containing 1.5% oregano EO against *Pseudomonas* spp. and lactic acid bacteria; Volpe et al. (2015) [30] of the coating based on 1% carrageenan incorporated with 1% lemon EO against H₂S producing bacteria and lactic acid bacteria; Yıldız & Yangılar (2016) [31] of coatings based on 8% whey protein concentrate/glycerol in ratios of 1:1 and 2:1 against lactic acid bacteria. On grass carp fillets, Yu et al. (2017) [32] have evaluated the efficacy of coatings based on 2% chitosan incorporated with different concentrations of glycerol monolaurate (0.1% and 0.3%) against *Pseudomonas* spp. and H₂S producing bacteria. In a study on pike-perch fillets, Shokri & Ehsani (2017) [33] have tested the efficacy of coatings based on 10% whey protein isolate incorporated with 2.5% lactoperoxidase, 1.5% and 3.0% α-tocopherol, respectively, combinations of lactoperoxidase and α-tocopherol (2.5%/1.5% and 2.5%/3.0%) against *Pseudomonas* spp. and H₂S producing bacteria.

Edible films/coatings incorporated with 2% thyme EO [28], 1.5% oregano EO [29], respectively 1% lemon EO [30] applied on rainbow trout fillets, 0.3% glycerol monolaurate [32] on grass carp fillets, and 2.5% lactoperoxidase [33] on pike-perch fillets have been proven to be the most effective against *Pseudomonas* spp. The most effective against H_2S producing bacteria were edible films/coatings incorporated with 2% thyme EO [28] applied on rainbow trout fillets, 0.3% glycerol monolaurate [32] on grass carp fillets, and 2.5% lactoperoxidase [33] on pike-perch fillets, but against lactic acid bacteria, the ones incorporated with 2% thyme EO [28], 1.5% oregano EO [29], 1% lemon EO [30], and 8% whey protein concentrate/glycerol, 2:1 [31] applied on rainbow trout fillets.

In a recent study, Carrión-Granda et al. (2018) [34] have examined the efficacy of coatings based on 10% whey protein isolate incorporated with different concentrations of oregano and thyme EOs (1% and 3%) under air and MAP conditions against *Pseudomonas* spp., H_2S producing bacteria, and lactic acid bacteria in hake fillets. The application of coating with 1% thyme EO under MAP has shown the best results against *Pseudomonas* spp. but against H_2S producing bacteria and lactic acid bacteria, the one with 3% oregano EO under the MAP. Different inhibitory effects displayed by an essential oil against various bacteria are most probably due to its chemical composition [35]. The antimicrobial mechanism of action of plant EOs is related to the hydrophobicity of their components [36], which enables them to migrate in the lipids of the bacterial cell membrane and mitochondria, disturbing their structures and rendering them more permeable [37]; leakage of ions and intracellular constituents can thus occur [38].

3.1.2. Efficacy against Pathogenic Microorganisms

According to current literature, few studies on the efficacy of active packaging materials against pathogenic microorganisms in fresh fish have been published. Findings of such in situ investigations are presented in Table 1. Gómez-Estaca et al. (2009) [39] have tested the efficacy of edible films based on 8% gelatin and 8% gelatin/chitosan, both incorporated with 7.5% clove EO on salmon fillets, in vitro against *Listeria innocua* and *Escherichia coli*, then in situ against total viable organisms. The film based on gelatin was more effective against both bacteria than the one based on gelatin/chitosan; the ionic and hydrogen bonds that were formed between gelatin and chitosan diminished the solubility of the resulting film, thus reducing the amount of clove EO released. However, in the in situ experiment, they used the film based on gelatin/chitosan for storage trials. Their previous studies revealed that the low water solubility of the gelatin/chitosan matrix gives the film stability under fish contact conditions during chilled storage.

There are also some studies on fish fillets challenged with pathogenic bacteria. Han et al. (2013) [40] have investigated the efficacy of films based on 6.75% (w/w) gelatin, with and without nisin-incorporated, against *Listeria monocytogenes* in rainbow trout fillets that were challenged with 2 log CFU/g inoculum before and after coating. The edible film incorporated with 18 µg/cm^2 nisin, applied before inoculation, showed the highest inhibitory effect on *Listeria monocytogenes*.

The efficacy of gelatin coatings containing different concentrations of oregano EO (0.5%, 1.0%, and 2.0% v/v) was also investigated by Min and Oh (2009) [41], in catfish fillets that were inoculated with *Salmonella typhimurium* and *Escherichia coli* O157:H7. The coating based on 3% (w/v) gelatin containing 2% oregano EO exhibited the best inhibitory effect on both bacteria.

3.1.3. Efficacy against Spoilage and/or Pathogenic Microorganisms

The following groups of microorganisms we have included into this category: total viable organisms, total mesophilic bacteria, total psychrotrophic bacteria, *Enterobacteriaceae* (including coliform bacteria), respectively total yeasts and moulds. Against total viable organisms, the most effective edible films/coatings were those that were incorporated with 2% thyme EO [28], 1.5% oregano EO [29], 1% lemon EO [30], and 1.5% cinnamon EO [42] applied on rainbow trout fillets, 0.3% glycerol monolaurate [32] on grass carp fillets, 2.5% lactoperoxidase [33] on pike-perch fillets, 0.2% tea polyphenols [43] on red drum fillets, 0.5% citric acid on Japanese sea bass fillets [44] and beluga sturgeon fillets [45], and 3% oregano EO under MAP conditions [34] on hake fillets. Edible coatings based on chitosan [46] applied to salmon fillets, respectively chitosan-gelatin [47] to golden pomfret fillets exhibited an antimicrobial effect compared to uncoated controls.

Regarding total psychrotrophic bacteria, the most effective were edible films/coatings incorporated with 2% thyme EO [28], 1.5% oregano EO [29], and 1.5% cinnamon EO [42] applied on rainbow trout fillets, 0.3% glycerol monolaurate [32] on grass carp fillets, 1.5% cinnamon EO on beluga sturgeon fillets [45], and 2.5% lactoperoxidase [33] on pike-perch fillets.

Edible coatings with 8% whey protein concentrate/glycerol, 2:1 applied on rainbow trout fillets [31], 1% chitosan [48] on salmon fillets, and 2% nanochitosan on silver carp fillets [49] have shown to be effective against both total psychrotrophic bacteria and total mesophilic bacteria.

The most effective edible films/coatings against *Enterobacteriaceae* (including coliform bacteria) were those incorporated with 2% thyme EO [28], 1.5% oregano EO [29], and 1% lemon EO [30] that were applied on rainbow trout fillets. Edible coating with 8% whey protein concentrate/glycerol, 2:1 has also shown to be effective against *Enterobacteriaceae* in rainbow trout fillets as compared with the other formulations tested in the study [31].

When tested against total yeasts and moulds, the edible coating based on 0.4% chitosan and 3.6% gelatin applied to golden pomfret fillets was the most effective among all formulations [47].

In the work of Carrión-Granda et al. (2018) [34], the edible coating incorporated with 3% oregano EO was the most effective against total viable organisms, total psychrotrophic bacteria, as well as *Enterobacteriaceae* when applied under the MAP conditions.

The results of these investigations are not comparable, since, on the same fish species, were applied edible films/coatings with different polymer matrices, respectively active agents and evaluated in different storage conditions (temperature, atmosphere composition, and storage time). We noticed, however, some tendencies that allow us to affirm that:

- edible films/coatings with the highest concentration of active agent tested have shown the greatest antimicrobial efficacy;
- antimicrobial films/coatings were more effective at lower temperatures when tested in different storage temperature conditions; and,
- under modified atmosphere packaging conditions, antimicrobial films/coatings were more effective than under air conditions.

Other authors have noticed that the effectiveness of antimicrobial packaging material depends also on the initial microbial load [40], chemical composition, and pH of tested food products [37]. Generally, the susceptibility of bacteria to the antimicrobial effect of EOs is increased in products with low-fat content and low pH, respectively.

Table 1. Antimicrobial films and coatings used for packaging fish.

Tested Fish Product	Antimicrobial Packaging Materials		Storage Conditions	Targeted Microorganism/ Group	Type of Microorganism	Level of Effectiveness against Targeted Microorganisms/Group at the End of Monitoring Time	MAL for Targeted Microorganism/ Group	Shelf-life of Fish Product		Ref.
	Film/Coating	Active Agent/ Concentration						Uncoated	Treated	
	Coating based on 2% (w/v) chitosan, acetic acid, and glycerol	Cinnamon EO/1.5% (v/v)	4 °C/16 days	Total viable organisms	Pathogenic and/or spoilage	1.5% (v/v) cinnamon EO > control	7.0 log CFU/g for TVC	Uncoated control-up to 8 days	Control-up to 16 days; 1.5% (v/v) cinnamon EO-up to 16 days	[42]
				Total psychrotrophic bacteria	Pathogenic and/or spoilage	Idem section TVC	7.0 log CFU/g for TPC	See section TVC	See section TVC	
Rainbow trout fillets	Film based on 1% (w/w) quince seed mucilage, glycerol, and Tween 80	Oregano EO/1%, 1.5%, and 2% (v/v) Thyme EO/1%, 1.5%, and 2% (v/v)	4 °C/18 days	Pseudomonas spp.	Spoilage	2% (v/v) thyme EO > 2% (v/v) oregano EO > 1.5 (v/v) thyme EO > 1.5% (v/v) oregano EO > 1% (v/v) thyme EO > 1% (v/v) oregano EO > control	7.0 log CFU/g for Pseuomonas spp.	See section TVC	See section TVC	[28]
				H$_2$S producing bacteria	Spoilage	2% (v/v) thyme EO > 2% (v/v) oregano EO > 1.5 (v/v) thyme EO > 1% (v/v) thyme EO > 1.5% (v/v) oregano EO > 1% (v/v) oregano EO > control	7.0 log CFU/g for H$_2$S producing bacteria	See section TVC	See section TVC	
				Lactic acid bacteria	Spoilage	2% (v/v) thyme EO > 1.5% (v/v) thyme EO > 1% (v/v) thyme EO > 2% (v/v) oregano EO > 1.5% (v/v) oregano EO > 1% (v/v) oregano EO > control	6.0 log CFU/g for LAB	See section TVC	See section TVC	
				Total viable organisms	Pathogenic and/or spoilage	2% (v/v) thyme EO > 1.5 (v/v) thyme EO > 2% (v/v) oregano EO > 1% (v/v) thyme EO > 1.5% (v/v) oregano EO > 1% (v/v) oregano EO > control	7.0 log CFU/g for TVC	Uncoated control-up to 6 days	Control-up to 9 days; 1% (v/v) Oregano EO-up to 9 days; 1.5% (v/v) Oregano EO-up to 12 days; 2% (v/v) Oregano EO-up to 15 days; 1% (v/v) Thyme EO-up to 12 days; 1.5% (v/v) Thyme EO-up to 15 days; 2% (v/v) Thyme EO-up to 18 days	
				Total psychrotrophic bacteria	Pathogenic and/or spoilage	2% (v/v) thyme EO > 1.5% (v/v) thyme EO > 2% (v/v) oregano EO > 1.5% (v/v) oregano EO > 1% (v/v) thyme EO > 1% (v/v) oregano EO > control	7.0 log CFU/g for TPC	See section TVC	See section TVC	
				Enterobacteriaceae	Pathogenic and/or spoilage	Idem section Pseudomonas spp.	5.0 log CFU/g for Enterobacteriaceae	See section TVC	See section TVC	
	Film based on 3% (w/v) gelatin and 1.5% (w/v) alginate, glycerol, and Tween 80	Oregano EO/1.5% (w/v)	4 °C/15 days	Pseudomonas spp.	Spoilage	1.5% (w/v) oregano EO > control	–	See section TVC	See section TVC	[29]
				Lactic acid bacteria	Spoilage	Idem section Pseudomonas spp.	–	See section TVC	See section TVC	
				Total viable organisms	Pathogenic and/or spoilage	Idem section Pseudomonas spp.	7.0 log CFU/g for TVC	Uncoated control-up to 3 days	Control-up to 3 days; 1.5% (w/v) oregano EO-up to 9 days	
				Total psychrotrophic bacteria	Pathogenic and/or spoilage	Idem section Pseudomonas spp.	–	See section TVC	See section TVC	
				Enterobacteriaceae	Pathogenic and/or spoilage	Idem section Pseudomonas spp.	–	See section TVC	See section TVC	

Table 1. *Cont.*

Tested Fish Product	Antimicrobial Packaging Materials		Storage Conditions	Targeted Microorganism/ Group	Type of Microorganism	Level of Effectiveness against Targeted Microorganisms/Group at the End of Monitoring Time	MAL for Targeted Microorganism/ Group	Shelf-life of Fish Product		Ref.
	Film/Coating	Active Agent/ Concentration						Uncoated	Treated	
	Coating based on 1% (w/w) carrageenan	Lemon EO, 1% (w/w)	4 °C/15 days	H$_2$S producing bacteria	Spoilage	1% (w/w) lemon EO > control	–	See section TVC	See section TVC	[30]
				Lactic acid bacteria	Spoilage	Idem section H$_2$S producing bacteria	–	See section TVC	See section TVC	
				Total viable organisms	Pathogenic and/or spoilage	Idem section H$_2$S producing bacteria	7.0 log CFU/g for TVC	Uncoated control-up to 3 days	Control-up to 12 days 1% (w/w) lemon EO-up to 15 days	
				Enterobacteriaceae	Pathogenic and/or spoilage	Idem section H$_2$S producing bacteria	–	See section TVC	See section TVC	
	Coating based on 8% (w/w) whey protein concentrate	–	4 °C/15 days	Lactic acid bacteria	Spoilage	8% (w/w) whey protein concentrate/glycerol, 2:1 > 8% (w/w) whey protein concentrate/glycerol, 1:1 > 8% (w/w) whey protein concentrate	–	See section TMC	See section TMC	[31]
				Total mesophilic bacteria	Pathogenic and/or spoilage	Idem section LAB	–	Uncoated control-up to 9 days	8% (w/w) whey protein concentrate-up to 12 days 8% (w/w) whey protein concentrate/glycerol, 1:1-up to 15 days 8% (w/w) whey protein concentrate/glycerol, 2:1-up to 15 days	
	Coating based on 8% (w/w) whey protein concentrate/glycerol, 1:1 and 2:1			Total psychrotrophic bacteria	Pathogenic and/or spoilage	Idem section LAB	–	See section TMC	See section TMC	
				Enterobacteriaceae	Pathogenic and/or spoilage	8% (w/w) whey protein concentrate/glycerol, 2:1 > 8% (w/w) whey protein concentrate > 8% (w/w) whey protein concentrate/glycerol, 1:1	–	See section TMC	See section TMC	
Silver carp fillets	Coating based on 2% (w/v) chitosan and glycerol	–	4 °C/12 days	Total mesophilic bacteria	Pathogenic and/or spoilage	2% (w/v) nanochitosan > 2% (w/v) chitosan	7.0 log CFU/g for TMC	See section TPC	See section TPC	[49]
	Coating based on 2% (w/v) nanochitosan and glycerol			Total psychrotrophic bacteria	Pathogenic and/or spoilage	Idem section TMC	7.0 log CFU/g for TPC	Uncoated control-up to 6 days1% glacial acetic acid-up to 6 days	2% (w/v) chitosan-up to 9 days 2% (w/v) nanochitosan-up to 12 days	

Table 1. *Cont.*

Tested Fish Product	Antimicrobial Packaging Materials		Storage Conditions	Targeted Microorganism/ Group	Type of Microorganism	Level of Effectiveness against Targeted Microorganisms/Group at the End of Monitoring Time	MAL for Targeted Microorganism/ Group	Shelf-life of Fish Product		Ref.
	Film/Coating	Active Agent/ Concentration						Uncoated	Treated	
Grass carp fillets	Coating based on 2% (*w*/*v*) chitosan, acetic acid, and glycerol	Glycerol monolaurate/0.1% and 0.3%	4 °C/20 days	*Pseudomonas* spp.	Spoilage	0.3% glycerol monolaurate > 0.1% glycerol monolaurate > control	–	See section TVC	See section TVC	[32]
				H$_2$S producing bacteria	Spoilage	Idem section *Pseudomonas* spp.	–	See section TVC	See section TVC	
				Total viable organisms	Pathogenic and/or spoilage	Idem section *Pseudomonas* spp.	7.0 log CFU/g for TVC	Uncoated control-up to 7 days	Control-up to 15 days / 0.1% glycerol monolaurate-up to 15 days / 0.3% glycerol monolaurate-up to 20 days	
				Total psychrotrophic bacteria	Pathogenic and/or spoilage	Idem section *Pseudomonas* spp.	–	See section TVC	See section TVC	
Beluga sturgeon fillets	Coating based on 8% (*w*/*v*) whey protein concentrate, glycerol, and Tween 80	Cinnamon EO/1.5% (*v*/*v*)	4 °C/20 days	Total viable organisms	Pathogenic and/or spoilage	1.5% (*v*/*v*) cinnamon EO > control	7.0 log CFU/g for TVC	Uncoated control-up to 4 days	Control-up to 4 days / 1.5% (*v*/*v*) cinnamon EO-up to 16 days / See section TVC	[45]
				Total psychrotrophic bacteria	Pathogenic and/or spoilage	Idem section TVC	–	See section TVC	See section TVC	
Salmon fillets	Coating based on 1% (*w*/*w*) chitosan, acetic acid, and glycerol	–	2 °C/6 days	Total mesophilic bacteria	Pathogenic and/or spoilage	1% (*w*/*w*) chitosan > 1% (*w*/*w*) chitosan and 2% (*w*/*w*) tapioca starch	–	Not specified	All treated samples-up to 6 days	[48]
	Coating based on 1% (*w*/*w*) chitosan, acetic acid, glycerol, and 2% (*w*/*w*) tapioca starch			Total psychrotrophic bacteria	Pathogenic and/or spoilage	Idem section TMC	–	See section TMC	See section TMC	
	Film based on 8% (*w*/*v*) gelatin/chitosan, 3:1, sorbitol and glycerol	Clove EO/7.5% (*v*/*w*)	2 °C/11 days	Total viable organisms	Pathogenic and/or spoilage	7.5% (*v*/*w*) clove EO	–	Uncoated control-up to 9 days	7.5% (*v*/*w*) clove EO-up to 11 days	[39]
	Coating based on 1.0, 1.5, and 2% (*w*/*v*) chitosan, lactic acid solution, and Tween 80	–	0 °C/18 days	Total viable organisms	Pathogenic and/or spoilage	1%, 1.5%, and 2% (*w*/*v*) chitosan	7.0 log CFU/g for TVC	Uncoated control-up to 9 days	All treated samples-up to 15 days	[46]

Table 1. *Cont.*

Tested Fish Product	Antimicrobial Packaging Materials		Storage Conditions	Targeted Microorganism/ Group	Type of Microorganism	Level of Effectiveness against Targeted Microorganisms/Group at the End of Monitoring Time	MAL for Targeted Microorganism/ Group	Shelf-life of Fish Product		Ref.
	Film/Coating	Active Agent/ Concentration						Uncoated	Treated	
Pike-perch fillets	Coating based on 10% (w/v) whey protein isolate, glycerol, and Tween 80	Lactoperoxidase/2.5% (v/v)	4 °C/16 days	Pseudomonas fluorescens	Spoilage	2.5% (v/v) lactoperoxidase > 2.5% (v/v) lactoperoxidase and 1.5% (v/v) α-tocopherol > 2.5% (v/v) lactoperoxidase and 3% (v/v) α-tocopherol > 3% (v/v) α-tocopherol > 1.5% (v/v) α-tocopherol > 10% (w/v) whey protein isolate > 10% (w/v) whey protein isolate and 3% (v/v) ethanol	–	See section TVC	See section TVC	[33]
				H₂S producing bacteria	Spoilage	Idem section *Pseudomonas fluorescens*	–	See section TVC	See section TVC	
	Coating based on 10% (w/v) whey protein isolate, glycerol, ethanol, and Tween 80	α-Tocopherol/1.5% (v/v) α-Tocopherol/3% (v/v) Lactoperoxidase and α-tocopherol/2.5% (v/v) and 1.5% (v/v) Lactoperoxidase and α-tocopherol/2.5% (v/v) and 3% (v/v)		Total viable organisms	Pathogenic and/or spoilage	2.5% (v/v) lactoperoxidase > 2.5% (v/v) lactoperoxidase and 1.5% (v/v) α-tocopherol > 2.5% (v/v) lactoperoxidase and 3% (v/v) α-tocopherol > 3% (v/v) α-tocopherol > 1.5% (v/v) α-tocopherol > control for coating with lactoperoxidase >control for other coatings	7.0 log CFU/g for TVC	– –	Control-up to 4 days 2.5% (v/v) lactoperoxidase-up to 12 days Control-up to 4 days 1.5% (v/v) α-tocopherol-up to 4 days 3% (v/v) α-tocopherol-up to 4 days 2.5% (v/v) lactoperoxidase and 1.5% (v/v) α-tocopherol-up to 12 days 2.5% (v/v) lactoperoxidase and 3% (v/v) α-tocopherol-up to 8 days	
				Total psychrotrophic bacteria	Pathogenic and/or spoilage	Idem section TVC	7.0 log CFU/g for TPC	See section TVC	See section TVC	
Japanese sea bass fillets	Coating based on 1.5% (w/v) chitosan and acetic acid	Citric acid/0.5% (v/v) Licorice extract/1% (v/v)	4 °C/12 days	Total viable organisms	Pathogenic and/or spoilage	0.5% (w/v) citric acid > 1% (w/v) licorice extract > control	6.0 log CFU/g for TVC	Uncoated control-up to 8 days	Control-up to 8 days 0.5% (w/v) citric acid-up to 12 days 1% licorice extract-up to 12 days	[44]
Red drum fillets	Coating based on 1.5% chitosan, acetic acid, and glycerol	Grape seed extract/0.2% (w/v) Tea polyphenols/0.2% (w/v)	4 °C/20 days	Total viable organisms	Pathogenic and/or spoilage	0.2% (w/v) tea polyphenols > 0.2% (w/v) grape seed extract	7.0 log CFU/g for TVC	Uncoated control-up to 8 days	0.2% (w/v) grape seed extract-up to 16 days 0.2% (w/v) tea polyphenols-up to 16 days	[43]

Table 1. *Cont.*

Tested Fish Product	Antimicrobial Packaging Materials		Storage Conditions	Targeted Microorganism/ Group	Type of Microorganism	Level of Effectiveness against Targeted Microorganisms/Group at the End of Monitoring Time	MAL for Targeted Microorganism/ Group	Shelf-life of Fish Product		Ref.
	Film/Coating	Active Agent/ Concentration						Uncoated	Treated	
Golden pomfret fillets	Coating based on 0.4% (*w*/*w*) chitosan	–	4 °C/17 days	Total viable organisms	Pathogenic and/or spoilage	0.4% (*w*/*w*) chitosan = 0.4% (*w*/*w*) chitosan and 3.6% (*w*/*w*) gelatin = 0.4% (*w*/*w*) chitosan and 5.4% (*w*/*w*) gelatin = 0.4% (*w*/*w*) chitosan and 7.2% (*w*/*w*) gelatin	6.0 log CFU/g for TVC	Deionized water-up to 17 days	All treated samples-up to 17 days	[47]
	Coating based on 0.4% (*w*/*w*) chitosan and gelatin			Total yeasts and moulds	Pathogenic and/or spoilage	0.4% (*w*/*w*) chitosan and 3.6% (*w*/*w*) gelatin > 0.4% (*w*/*w*) chitosan and 5.4% (*w*/*w*) gelatin > 0.4% (*w*/*w*) chitosan and 7.2% (*w*/*w*) gelatin > 0.4% (*w*/*w*) chitosan	–	See section TVC	See section TVC	
Hake fillets	Coating based on 10% (*w*/*w*) whey protein isolate and glycerol	Oregano EO/1% and 3% (*w*/*w*) Thyme EO/1% and 3% (*w*/*w*)	4 °C/8 days	Total viable organisms	Pathogenic and/or spoilage	3% (*w*/*w*) thyme EO > 1% (*w*/*w*) thyme EO > 3% (*w*/*w*) oregano EO > 1% (*w*/*w*) oregano EO > control	7.0 log CFU/g for TVC	Uncoated control-up to 4 days	All treated samples-up to 4 days	[34]
				Total psychrotrophic bacteria	Pathogenic and/or spoilage	Idem section TVC	7.0 log CFU/g for TPC	See section TVC	See section TVC	
				Enterobacteriaceae	Pathogenic and/or spoilage	3% (*w*/*w*) oregano EO > 3% (*w*/*w*) thyme EO >1% (*w*/*w*) thyme EO > 1% (*w*/*w*) oregano EO > control	4.0 log CFU/g for *Enterobacteriaceae*	See section TVC	See section TVC	
				Lactic acid bacteria	Spoilage	Idem section *Enterobacteriaceae*	–	See section TVC	See section TVC	
				H$_2$S producing bacteria	Spoilage	Idem section TVC	–	See section TVC	See section TVC	
				Pseudomonas spp.	Spoilage	Idem section TVC	–	See section TVC	See section TVC	
			4 °C under MAP conditions/ 16 days	Total viable organisms	Pathogenic and/or spoilage	3% (*w*/*w*) thyme EO > 3% (*w*/*w*) oregano EO > 1% (*w*/*w*) thyme EO > 1% (*w*/*w*) oregano EO > control	7.0 log CFU/g for TVC	Uncoated control-up to 8 day	Control-up to 8 days 3% (*w*/*w*) oregano EO-up to 16 days 1% (*w*/*w*) oregano EO-up to 8 days 3% (*w*/*w*) thyme EO-up to 16 days 1% (*w*/*w*) thyme EO-up to 16 days	
				Total psychrotrophic bacteria	Pathogenic and/or spoilage	3% (*w*/*w*) oregano EO > 1% (*w*/*w*) thyme EO > 3% (*w*/*w*) thyme EO > 1% (*w*/*w*) oregano EO > control	7.0 log CFU/g for TPC	See section TVC	See section TVC	
				Enterobacteriaceae	Pathogenic and/or spoilage	3% (*w*/*w*) oregano EO > 3% (*w*/*w*) thyme EO > 1% (*w*/*w*) thyme EO > 1% (*w*/*w*) oregano EO > control	4.0 log CFU/g for *Enterobacteriaceae*	See section TVC	See section TVC	
				Lactic acid bacteria	Spoilage	3% (*w*/*w*) oregano EO >3% (*w*/*w*) thyme EO > 1% (*w*/*w*) oregano EO >1% (*w*/*w*) thyme EO > control	–	See section TVC	See section TVC	
				H$_2$S producing bacteria	Spoilage	Idem section LAB	–	See section TVC	See section TVC	
				Pseudomonas spp.	Spoilage	1% (*w*/*w*) thyme EO > 3% (*w*/*w*) oregano EO > 1% (*w*/*w*) oregano EO > 3% (*w*/*w*) thyme EO > control	–	See section TVC	See section TVC	
		Oregano EO/1% and 3% (*w*/*w*)	4 °C under air and MAP conditions/ 12 days	Total viable organisms	Pathogenic and/or spoilage	3% (*w*/*w*) oregano EO (MAP) > 3% (*w*/*w*) oregano EO (air)	7.0 log CFU/g for TVC	Uncoated control (air)-up to 4 daysUncoated (MAP)-up to 4 days	3% (*w*/*w*) oregano EO (MAP)-up to 12 days 3% (*w*/*w*) oregano EO (air)-up to 4 days	
				Total psychrotrophic bacteria	Pathogenic and/or spoilage	Idem section TVC	7.0 log CFU/g for TVC	See section TVC	See section TVC	
				Enterobacteriaceae	Pathogenic and/or spoilage	3% (*w*/*w*) oregano EO (MAP) > 3% (*w*/*w*) oregano EO (air)	4.0 log CFU/g for *Enterobacteriaceae*	See section TVC	See section TVC	
				Lactic acid bacteria	Spoilage	Idem section *Enterobacteriaceae*	–	See section TVC	See section TVC	
				H$_2$S producing bacteria	Spoilage	Idem section *Enterobacteriaceae*	–	See section TVC	See section TVC	
				Pseudomonas spp.	Spoilage	Idem section *Enterobacteriaceae*	–	See section TVC	See section TVC	

EO, essential oil; CFU, colony-forming units; TVC, total viable count; TMC, total mesophilic bacteria; TPC, total psychrotrophic bacteria; LAB, Lactic Acid Bacteria; MAL, maximum acceptable level.

3.2. Efficacy of Edible Films/Coatings on Enhancing the Shelf-Life of Fresh Fish

The application of above-mentioned edible films and coatings to fish fillets resulted in an extension of their shelf-life as compared to uncoated controls. The film based on 1% quince seed mucilage incorporated with 2% thyme EO prolonged the shelf-life of rainbow trout fillets by 12 days [28] and the one based on 3% gelatin and 1.5% alginate incorporating 1.5% oregano EO by 6 days [29]; the coating based on 2% chitosan incorporated with 1.5% cinnamon EO by 8 days [42], the one based on 1% carrageenan incorporated with 1% lemon EO by 12 days [30], and the one based on 8% whey protein concentrate/glycerol, 2:1 by 6 days [31]. In these cases, the shelf-life was stated considering a maximum acceptable level of 7.0 log CFU/g for the total viable count.

Shelf-lives of silver carp and grass carp fillets were extended by 6 and 13 days, respectively, when coatings based on 2% nanochitosan [49] and 2% chitosan incorporated with 0.3% glycerol monolaurate [32] was used.

When applied to salmon fillets, the film based on 8% gelatin/chitosan, 3:1 incorporated with 7.5% clove EO [39] and coatings based on 1%, 1.5%, and 2% chitosan [46] enhanced the shelf-lives by 6 days.

On beluga sturgeon fillets, the coating based on 8% whey protein concentrate incorporated with 1.5% cinnamon EO [45] extended the shelf-life by 12 days.

The study of Shokri & Ehsani (2017) [33] on pike-perch fillets show a shelf-life prolongation by 8 days when a packaging material based on 10% whey protein isolate incorporated with 2.5% lactoperoxidase was used for coating.

Another study, carried out by Qiu et al. (2014) [44], has shown an increased storage stability (from 8 to 12 days) of Japanese sea bass fillets coated with a solution containing 1.5% chitosan and 0.5% citric acid [44].

The coating formulation of Li et al. (2013) [43], also based on 1.5% chitosan but incorporated with 0.2% tea polyphenols, prolonged the microbiological shelf-life of red drum fillets by 8 days.

In a study on hake fillets, Carrión-Granda et al. (2018) [34] reported a shelf-life prolongation by 8 days when a coating based on 10% whey protein isolate incorporated with 3% oregano EO was used under MAP conditions.

Our review also revealed some studies in the existing literature focused on the application of synthetic films to fresh fish fillets. Cardoso et al. (2017) [50] have tested the efficiency of films based on poly(butylene adipate-co-terephthalate) incorporated with different levels of oregano EO (2.5%, 5.0%, 7.5%, and 10%) in lessening coliform bacteria, *Staphylococcus aureus*, and total psychrotrophic bacteria in fish fillets. The film incorporated with 10% (w/w) oregano EO showed the highest inhibitory effect on all bacteria leading to a shelf-life extension of 6 days for wrapped samples. The shelf-life was established considering a maximum acceptable level of 5.0 log CFU/g for *Staphylococcus aureus*.

In another study, Rollini et al. (2016) [51] have evaluated the efficacy of film based on polyethylene terephthalate coated with 3% (w/v) lysozyme and lactoferrin water solution, respectively, coextruded multilayer film based on polypropylene incorporated with 4.8% carvacrol against total mesophilic bacteria, total psychrotrophic bacteria, *Enterobacteriaceae* (including coliform bacteria), lactic acid bacteria, *Pseudomonas* spp., and H_2S producing bacteria. The film that was coated with 3% lysozyme-lactoferrin has shown the best antibacterial results on total mesophilic bacteria, total psychrotrophic bacteria, lactic acid bacteria, and H_2S producing bacteria, but the one incorporated with 4.8% carvacrol on *Enterobacteriaceae* (including coliform bacteria) and *Pseudomonas* spp. All of the samples were stored for up to four days; therefore, no extension of shelf-life was possible to notice for treated samples in such a short period of storage.

At high levels of incorporation with EOs, active films/coatings may impart foreign flavours to the products on which are applied. Of all the studies that are mentioned in Table 1, only two mentioned their effects on the sensory attributes of fresh fish. The study of Jouki et al. (2014) [28] revealed no significant negative effect of films based on 1% quince seed mucilage incorporated with oregano and thyme EOs in concentrations of up to 2% on the organoleptic acceptability of rainbow trout fillets.

Similar observations were also reported by Ojagh et al. (2010) [42] when a coating based on 2% chitosan incorporated with 1.5% cinnamon EO treatment was applied.

3.3. Effects of Edible Films/Coatings on the Chemical Quality of Fresh Fish

Table 2 summarizes the effects of the above-mentioned edible films and coatings on the chemical quality of fresh fish. Chemical indicators of lipid oxidation (TBARS—thiobarbituric acid reactive substances), degradation of nitrogen-containing compounds (TVB-N—total volatile basic nitrogen and TMA-N—trimethylamine nitrogen), and adenosine triphosphate breakdown (k-value) were measured during storage of fish fillets.

The thiobarbituric acid reactive substances (TBARS) assay is commonly used to evaluate malondialdehyde (MDA) content. MDA is one of the most significant products of lipid damage [52]. Several researchers [28,33,42] have proposed maximum permitted levels for TBARS although the threshold criteria have not yet received regulatory approval; values <3 mg MDA/kg for perfect quality material, $3 \leq$ MDA/kg < 5 for good quality material, and $5 \leq$ MDA/kg < 8 for suitable for human consumption. In the published data reviewed in the current paper, TBARS values ranged from 0.2 to 0.9 mg MDA/kg for rainbow trout fillets, 3.0 to 4.0 mg MDA/kg for silver carp fillets, 0.9 to 1.2 mg MDA/kg for grass carp fillets, 0.06 to 0.12 mg MDA/kg for beluga sturgeon fillets, 1.1 to 1.8 mg MDA/kg for salmon fillets, 1.0 to 2.5 mg MDA/kg for pike-perch fillets, 0.2 to 2.0 mg MDA/kg for Japanese sea bass fillets, and 0.8 to 1.8 mg MDA/kg for red drum fillets; samples meeting the requirements for good quality material, respectively perfect quality material.

Total volatile base nitrogen (TVB-N) is one of the most widely used fish spoilage indicator [53]. It represents the sum of ammonia, methylamine, dimethylamine, trimethylamine, and other basic nitrogenous volatile compounds resulted from fish degradation [54,55]. Commission Regulation (EC) 2074/2005 [56] set limits for TVB-N only for redfish, flatfish, Atlantic salmon, hake, and gadoids; values $\leq$25 mg N/100 g for *Sebastes* spp., *Helicolenus dactylopterus*, and *Sebastichthys capensis*, $\leq$30 mg N/100 g for species belonging to the *Pleuronectidae* family (with the exception of halibut: *Hippoglossus* spp.), and $\leq$35 mg N/100 g for *Salmo salar*, species belonging to the *Merlucciidae* family, and species belonging to the *Gadidae* family. Since no limits of acceptability for rainbow trout, grass carp, beluga sturgeon, pike-perch, Japanese sea bass, and red drum have been established by EC Regulation 2074/2005 [56], the values that were reported previously in the literature were taken as threshold limits by Ojagh et al. (2010) [42], Jouki et al. (2014) [28], Kazemi & Rezaei (2015) [29], Volpe et al. (2015) [30], Yıldız & Yangılar (2016) [31], Yu et al. (2017) [32], Bahram et al. (2016) [45], Shokri and Ehsani (2017) [33], Qiu et al. (2014) [44], and Li et al. (2013) [43]; levels of 25–35 mg N/100 g for rainbow trout, $\leq$15 mg N/100 for grass carp, levels of 35–40 mg N/100 g for beluga sturgeon, $\leq$35 mg N/100 for pike-perch, levels of 30–35 mg N/100 g for Japanese sea bass, and $\leq$25 mg N/100 for red drum. TVB-N values reported in the reviewed studies ranged from 10 to 65 mg N/100 g for rainbow trout fillets, 44 to 60 mg N/100 g for silver carp fillets, 15 to 28 mg N/100 g for grass carp fillets, 50 to 70 mg N/100 g for beluga sturgeon fillets, 28 to 33 mg N/100 g for salmon fillets, 35 to 45 mg N/100 g for pike-perch fillets, 30 to 100 mg N/100 g for Japanese sea bass fillets, 34 to 51 mg N/100 g for red drum fillets, and 11 to 94 mg N/100 g for golden pomfret fillets.

Most marine fish contain TMAO [57]. TMAO is also found, with few exceptions, in freshwater fish, but only in small concentrations [58]. Certain bacteria that occur naturally on the skin, in the guts of fish, and in water can break down TMAO to TMA. The amount of trimethylamine nitrogen (TMA-N) produced is a measure of the activity of spoilage bacteria in the flesh and so is an indicator of the degree of spoilage [57]. There are no regulatory limits available for TMA level in fish. The rejection limit proposed by Jouki et al. (2014) [28] was <5 mg N/100 g and by Souza et al. (2010) [46] $\leq$5 mg N/100 g.

Table 2. Effects of antimicrobial packaging on chemical quality of fresh fish.

Tested Fish Product	Antimicrobial Packaging Material		Storage Conditions	ML Obtained for TBARS during Storage	TLV for TBA	ML Obtained for TVB-N during Storage	TLV for TVB-N	ML Obtained for TMA-N during Storage	TLV for TMA-N	ML Obtained for K-Value during Storage	TL for K-Value	Ref.
	Film/Coating	Active Agent/ Concentration										
Rainbow trout fillets	Coating based on 2% (w/v) chitosan, acetic acid, and glycerol	Cinnamon EO/1.5% (v/v)	4 °C/16 days	1.5% (v/v) cinnamon EO (~0.2 mg MDA/kg) < uncoated control (below 0.25 mg MDA/kg) < control (below 0.25 mg MDA/kg)	5 mg MDA/kg-good quality; 8 mg MDA/kg-suitable for human consumption	1.5% (v/v) cinnamon EO (~10 mg N/100 g) < control (~20 mg N/100 g) < uncoated control (~40 mg N/100 g)	25 mg N/100 g	–	–	–	–	[42]
	Film based or 1% (w/w) quince seed mucilage, glycerol, and Tween 80	Oregano EO/1%, 1.5%, and 2% (v/v) Thyme EO/1%, 1.5%, and 2% (v/v)	4 °C/18 days	2% (v/v) oregano EO (~0.4 mg MDA/kg) < 1.5% (v/v) oregano EO (~0.4 mg MDA/kg) < % (v/v) thyme EO (below 0.5 mg MDA/kg) < 1% (v/v) oregano EO (below 0.5 mg MDA/kg) < 1.5 (v/v) thyme EO (below 0.6 mg MDA/kg) < 1% (v/v) thyme EO (below 0.6 mg MDA/kg) < control (~0.8 mg MDA/kg) < uncoated control (~0.9 mg MDA/kg)	below 5 mg MDA/kg	2% (v/v) thyme EO (below 20 mg N/100 g) < 1.5 (v/v) thyme EO (below 25 mg N/100 g) < 2% (v/v) oregano EO (below 25 mg N/100 g) < 1% (v/v) thyme EO (below 30 mg N/100 g) < 1.5% (v/v) oregano EO (below 30 mg N/100 g) < 1% (v/v) oregano EO (below 35 mg N/100 g) < control (below 35 mg N/100 g) < uncoated control (below 45 mg N/100 g)	25 mg N/100 g	2% (v/v) thyme EO (~5 mg N/100 g) < 1.5% (v/v) thyme EO (~6 mg N/100 g) < 2% (v/v) oregano EO (~6 mg N/100 g) < 1% (v/v) thyme EO (below 8 mg N/100 g) < 1.5% (v/v) oregano EO (below 8 mg N/100 g) < 1% (v/v) oregano EO (~8 mg N/100 g) < uncoated control (~12 mg N/100 g) < control (~12 mg N/100 g)	below 5 mg N/100 g	–	–	[28]
	Film based on 2% (w/v) gelatin and 1.5% (v/v) alginate, glycerol, and Tween 80	Oregano EO/1.5% (w/v)	4 °C/15 days	–	–	1.5% (w/v) oregano EO (~60 mg N/100 g) < control (~65 mg N/100 g) < uncoated control (~65 mg N/100 g)	35 mg N/100 g	–	–	–	–	[29]
	Coating based on 1% (w/w) carrageenan	Lemon EO/1% (w/w)	4 °C/15 days	–	–	1% (w/w) lemon EO (20 mg N/100 g) < control (below 35 mg N/100 g) < uncoated control (40 mg N/100 g)	25 mg N/100 g	–	–	–	–	[30]
	Coating based on 8% (w/w) whey protein concentrate Coating based on 8% (w/w) whey protein concentrate/glycerol, 1:1 and 2:1	–	4 °C/15 days	8% (w/w) whey protein concentrate/glycerol, 2:1 (0.4 mg MDA/kg) < 8% (w/w) whey protein concentrate/glycerol, 1:1 (0.5 mg MDA/kg) < 8% (w/w) whey protein concentrate (0.6 mg MDA/kg) < uncoated control (0.7 mg MDA/kg)	–	8% (w/w) whey protein concentrate/glycerol, 2:1 (21.1 mg N/100 g) < 8% (w/w) whey protein concentrate/glycerol, 1:1 (24.6 mg N/100 g) < 8% (w/w) whey protein concentrate (27.4 mg N/100 g) < uncoated control (32.5 mg N/100 g)	25 mg N/100 g	–	–	–	–	[31]

Table 2. *Cont.*

Tested Fish Product	Antimicrobial Packaging Materials		Storage Conditions	ML Obtained for TBARS during Storage	TLV for TBA	ML Obtained for TVB-N during Storage	TLV for TVB-N	ML Obtained for TMA-N during Storage	TLV for TMA-N	ML Obtained for *K*-Value during Storage	TL for *K*-Value	Ref.
	Film/Coating	Active Agent/ Concentration										
Silver carp fillets	Coating based on 2% (*w*/*v*) chitosan and glycerolCoating based on 2% (*w*/*v*) nanochitosan and glycerol	–	4 °C/12 days	2% (*w*/*v*) chitosan (below 3 mg MDA/kg) < 2% (*w*/*v*) nanochitosan (below 3 mg MDA/kg) < uncoated control (below 4 mg MDA/kg) < 1% glacial acetic acid (~4 mg MDA/kg)	–	2% (*w*/*v*) nanochitosan (44.4 mg N/100 g) < 2% (*w*/*v*) chitosan (30.8 mg N/100 g) < 1% glacial acetic acid (below 60 mg N/100 g) < uncoated control (~60 mg N/100 g)	–	–	–	–	–	[49]
Grass carp fillets	Coating based on 2% (*w*/*v*) chitosan, acetic acid, and glycerol	Glycerol monolaurate/ 0.1% and 0.3%	4 °C/ 20 days	0.3% glycerol monolaurate (~0.9 mg MDA/kg) < 0.1% glycerol monolaurate (~0.9 mg MDA/kg) < ontrol (~0.9 mg MDA/kg) < uncoated control (below 1.2 mg MDA/kg)	–	0.3% glycerol monolaurate (15 mg N/100 g) < 0.1% glycerol monolaurate (below 20 mg N/100 g) < control (~22.5 mg N/100 g) < uncoated control (~27.5 mg N/100 g)	15 mg N/100 g	–	–	0.3% glycerol monolaurate (~69%) < 0.1% glycerol monolaurate (77.7%) < control (78.2%) < uncoated control (90.5%)	<20%-vf; <60%-mf; >60%-rp	[32]
Beluga sturgeon fillets	Coating based on 8% (*w*/*v*) whey protein concentrate, glycerol, and Tween 80	Cinnamon EO/1.5% (*v*/*v*)	4%°C/ 20 days	1.5% (*v*/*v*) cinnamon EO (below 0.06 mg MDA/kg) < control (below 0.1 mg MDA/kg) < uncoated control (below 0.12 mg MDA/kg)	–	1.5% (*v*/*v*) cinnamon EO (~50 mg N/100 g) < control (below 70 mg N/100 g) < uncoated control (~70 mg N/100 g)	35–40 mg N/100 g	–	–	–	–	[45]
Salmon fillets	Coating based on 1.0%, 1.5%, and 2% (*w*/*v*) chitosan, lactic acid solution, and Tween 80	–	0 °C/ 18 days	All treated samples (1.1 mg MDA/kg) < uncoated control (1.8 mg MDA/kg)	1 mg MDA/kg	All treated samples (28 mg N/100 g) < uncoated control (33 mg N/100 g)	30 mg TVB-N/100 g	All treated samples (5 mg N/100 g) < uncoated control (6 mg N/100 g)	5 mg N/100 g	All treated samples (46%) < uncoated control (50%)	40%	[46]
Pike-perch fillets	Coating based on 10% (*w*/*v*) whey protein isolate, glycerol, and Tween 80	Lactoperoxidase/ 2.5% (*v*/*v*)	4 °C/ 16 days	3% (*v*/*v*) α-tocopherol (below 1 mg MDA/kg) < 2.5% (*v*/*v*) lactoperoxidase and 3% (*v*/*v*) α-tocopherol (below 1 mg MDA/kg) < 1.5% (*v*/*v*) α-tocopherol (~1 mg MDA/kg) < 2.5% (*v*/*v*) lactoperoxidase and 1.5% (*v*/*v*) α-tocopherol (~1 mg MDA/kg) < control for other coatings (below 2.5 mg MDA/kg) < control for coating with lactoperoxidase (below 2.5 mg MDA/kg) < 2.5% (*v*/*v*) lactoperoxidase (~2.5 mg MDA/kg)	below 3 mg MDA/kg-perfect quality material; below 5 mg MDA/kg-good quality material	2.5% (*v*/*v*) lactoperoxidase (below 35 mg N/100 g) < 2.5% (*v*/*v*) lactoperoxidase and 1.5% (*v*/*v*) α-tocopherol (below 40 mg N/100 g) < control for coating with lactoperoxidase (~40 mg N/100 g) < control for other coatingl (~40 mg N/100 g) < 3% (*v*/*v*) α-tocopherol (below 45 mg N/100 g) < 1.5% (*v*/*v*) α-tocopherol (below 45 mg N/100 g) < 2.5% (*v*/*v*) lactoperoxidase and 3% (*v*/*v*) α-tocopherol (below 45 mg N/100 g)	35 mg N/100 g	–	–	–	–	[33]
	Coating based on 10% (*w*/*v*) whey protein isolate, glycerol, ethanol, and Tween 80	α-Tocopherol/ 1.5% (*v*/*v*) α-Tocopherol/ 3% (*v*/*v*) Lactoperoxidase and α-tocopherol/ 2.5% (*v*/*v*) and 1.5% (*v*/*v*) Lactoperoxidase and α-tocopherol/ 2.5% (*v*/*v*) and 3% (*v*/*v*)											

Table 2. *Cont.*

Tested Fish Product	Antimicrobial Packaging Materials		Storage Conditions	ML Obtained for TBARS during Storage	TLV for TBA	ML Obtained for TVB-N during Storage	TLV for TVB-N	ML Obtained for TMA-N during Storage	TLV for TMA-N	ML Obtained for K-Value during Storage	TL for K-Value	Ref.
	Film/Coating	Active Agent/ Concentration										
Japanese sea bass fillets	Coating based on 1.5% (*w/v*) chitosan and acetic acid	Citric acid/0.5% (*w/v*) Licorice extract/1% (*w/v*)	4 °C/ 12 days	0.5% (*w/v*) citric acid (~0.2 mg MDA/kg) < % (*w/v*) licorice extract (~0.2 mg MDA/kg) < control (below 1.5 mg MDA/kg) < uncoated control (below 2.0 mg MDA/kg)	–	0.5% (*w/v*) citric acid (29.7 mg N/100 g) < 1% (*w/v*) licorice extract (48.0 mg N/100 g) < control (60.5 mg N/100 g) < uncoated control (100.2 mg N/100 g)	30–35 mg N/100 g	–	–	–	–	[44]
Red drum fillets	Coating based on 1.5% chitosan, acetic acid, and glycerol	Grape seed extract/0.2% (*w/v*) Tea polyphenols/0.2% (*w/v*)	4 °C/20 days	0.2% (*w/v*) tea polyphenols (~0.8 mg MDA/kg) < 0.2% (*w/v*) grape seed extract (~1.0 mg MDA/kg) < uncoated control (~1.8 mg MDA/kg)	–	0.2% (*w/v*) tea polyphenols (33.69 mg N/100 g) < 0.2% (*w/v*) grape seed extract (38.17 mg N/100 g) < uncoated control (51.25 mg N/100 g)	25 mg N/100 g	–	–	0.2% (*w/v*) tea polyphenols (~40%) < 0.2% (*w/v*) grape seed extract (~45%) < uncoated control (62.57%)	60%	[43]
Golden pomfret fillets	Coating based on 0.4% (*w/w*) chitosan Coating based on 0.4% (*w/w*) chitosan and gelatin	–	4 °C/ 17 days	–	–	0.4% (*w/w*) chitosan and 7.2% (*w/w*) gelatin (10.51 mg N/100 g) < 0.4% (*w/w*) chitosan and 5.4% (*w/w*) gelatin (12.31 mg N/100 g) < 0.4% (*w/w*) chitosan and 3.6% (*w/w*) gelatin (13.48 mg N/100 g) < deionized water (93.52 mg N/100 g)	–	–	–	–	–	[47]

TBARS, thiobarbituric acid reactive substances; TLV, threshold limit value; ML, maximum levels; TVB-N, total volatile basic nitrogen; TMA-N, trimethylamine nitrogen; MDA, malondialdehyde; vf, very fresh; mf, moderately fresh; rp, rejection point.

K-value is an important chemical index widely used for fish freshness [59]. During post-mortem storage of fish, autolytic changes take place in the muscle that determines adenosine triphosphate (ATP) degradation with the formation of adenosine-5′-diphosphate (ADP), adenosine-5′-monophosphate (AMP), inosine-5′-monophosphate (IMP), inosine (HxR), and hypoxanthine (Hx). *K*-value is calculated as the percentage of the sum of HxR and Hx, divided by the sum of ATP, ADP, AMP, IMP, HxR, and Hx [12,59]. Since there are no legally enforceable limits for *k*-value in fish, Yu et al. (2017) [32] proposed the following freshness criteria: very fresh fish (*k*-value < 20%), moderately fresh (*k*-value < 60%), and spoiled (*k*-value > 60%). *K*-values reported in the discussed studies ranged from 68.7% to 90.5% for grass carp fillets, 46% to 50% for salmon fillets and 40% to 62.6% for red drum fillets; samples meeting freshness criteria for moderately fresh, respectively spoiled.

4. Conclusions

The active packaging of fish represents an economic alternative to conventional preservation technologies (vacuum and modified atmosphere packaging) due to the limited capital investment as compared to those. Besides being biodegradable, edible films and coatings improve the microbiological stability of fish and reduce waste; moreover, retard lipid oxidation. For the past 10 years, research on the use of antimicrobial packaging materials for fresh fish applications has undergone considerable evolution; nevertheless, as far as we know, there is not yet an edible film or coating commercially available on the market.

Fish represent one of the most-traded segments of the world food sector. Therefore, there is a great demand for the packaging of this good. Industrial production and commercialization of antimicrobial packaging materials for fresh fish could be an exploitable sector by the packaging industry. Suppliers of active packaging materials on the European market need to make sure that their products comply with the requirements of Regulations (EC) 1935/2004 [60] and (EC) 450/2009 [61] regarding active and intelligent materials that are intended to come into contact with food, respectively, Regulation (EC) 1333/2008 [62] that lays down specifications for food additives. Additional studies are however needed to further validate these findings, especially on the stability of antimicrobial films/coatings during shipment, storage, and handling.

Author Contributions: Conceptualization, C.A.S. and D.C.V.; Formal Analysis, M.-I.S.; Funding Acquisition, D.C.V.; Methodology, C.A.S. and D.C.V.; Supervision, C.A.S. and D.C.V.; Writing–Original Draft, M.-I.S.; Writing–Review & Editing, C.A.S. and D.C.V.

Funding: This research was funded by a grant of the Romanian National Authority for Scientific Research and Innovation, CCDI-UEFISCDI, project number 27/2018 CO FUND–MANUNET III-NON-ACT-2, within PNCDI III.

Acknowledgments: We are grateful for the administrative and financial support received from the University of Agricultural Sciences and Veterinary Medicine Cluj-Napoca, Romania.

Conflicts of Interest: The authors declare no conflict of interest.

References

1. FAO. *The State of World Fisheries and Aquaculture 2016: Contributing to Food Security and Nutrition for All*; FAO: Rome, Italy, 2016.
2. Watterson, A.; Little, D.; Young, J.A.; Boyd, K.; Azim, E.; Murray, F. Towards integration of environmental and health impact assessments for wild capture fishing and farmed fish with particular reference to public health and occupational health dimensions. *Int. J. Environ. Res. Public Health* **2008**, *5*, 258–277. [CrossRef] [PubMed]
3. Gustavsson, J.; Cederberg, C.; Sonesson, U.; van Otterdijk, R.; Meybeck, A. *Global Food Losses and Food Waste. Extent, Causes and Prevention*; FAO: Rome, Italy, 2011; pp. 10–14.
4. Akande, G.; Diei-Ouadi, Y. *Post-Harvest Losses in Small-Scale Fisheries: Case Studies in Five Sub-Saharan African Countries*; FAO Fisheries and Aquaculture Technical Paper No. 550; FAO: Rome, Italy, 2010; pp. XI–XIII.
5. Hall, G.M. *Fish Processing: Sustainability and New Opportunities*; Wiley-Blackwell: Oxford, UK, 2011; pp. 51–76.

6. Blackburn, C.D.W. Managing microbial food spoilage: an overview. In *Food Spoilage Microorganisms*, 1st ed.; Blackburn, C.D.W., Ed.; Woodhead Publishing Ltd.: Cambridge, UK, 2006; pp. 147–170.

7. Reducing Post-Harvest Losses. Available online: http://www.oceansatlas.org/subtopic/en/c/1337/ (accessed on 2 August 2018).

8. Makawa, Z.; Kapute, F.; Valeta, J. Effect of delayed processing on nutrient composition, pH and organoleptic quality of pond raised tilapia (*Oreochromis shiranus*) stored at ambient temperature. *Afr. J. Food Agric. Nutr. Dev.* **2014**, *14*, 8872–8884.

9. Tesfay, S.; Teferi, M. Assessment of fish post-harvest losses in Tekeze dam and Lake Hashenge fishery associations: northern Ethiopia. *Agric. Food Secur.* **2017**, *6*, 4. [CrossRef]

10. Calligaris, S.; Manzocco, L. Critical indicators in shelf life assessment. In *Shelf Life Assessment of Food*; Nicoli, M.C., Ed.; CRC Press/Taylor and Francis Group: Boca Raton, FL, USA, 2012; pp. 61–74.

11. Morrison, C.R. Fish and shellfish. In *Frozen Food Technology*; Mallet, C.P., Ed.; Blackie Academic & Professional: Glasgow, UK, 1993; pp. 196–236.

12. Huss, H.H. *Quality and Quality Changes in Fresh Fish*; FAO Fisheries Technical Paper No. 348; FAO: Rome, Italy, 1995; pp. 35–67.

13. Diei-Ouadi, Y.; Mgawe, Y.I. *Post-Harvest Fish Loss Assessment in Small-Scale Fisheries: A Guide for the Extension Officer*; FAO Fisheries and Aquaculture Technical Paper No. 559; FAO: Rome, Italy, 2011; pp. 3–11.

14. Reddy, N.R.; Villanueva, M.; Kautter, D.A. Shelf life of modified-atmosphere-packaged fresh tilapia fillets stored under refrigeration and temperature-abuse conditions. *J. Food Prot.* **1995**, *58*, 908–914. [CrossRef]

15. Getu, A.; Misganaw, K.; Bazezew, M. Post-harvesting and major related problems of fish production. *Fish. Aquac. J.* **2015**, *6*, 4. [CrossRef]

16. Ashie, I.N.A.; Smith, J.P.; Simpson, B.K. Spoilage and shelf-life extension of fresh fish and shellfish. *Crit. Rev. Food Sci. Nutr.* **1996**, *36*, 87–121. [CrossRef] [PubMed]

17. Vodnar, D.C.; Pop, O.L.; Dulf, F.V.; Socaciu, C. Antimicrobial efficiency of edible films in food industry. *Not. Bot. Horti Agrobo. Cluj-Napoca* **2015**, *43*, 302–312. [CrossRef]

18. Yildirim, S.; Röcker, B.; Pettersen, M.K.; Nilsen-Nygaard, J.; Ayhan, Z.; Rutkaite, R.; Radusin, T.; Suminska, P.; Marcos, B.; Coma, V. Active packaging applications for food. *Compr. Rev. Food Sci. Food Saf.* **2018**, *17*, 165–199. [CrossRef]

19. Velu, S.; Abu Bakar, F.; Mahyudin, N.A.; Saari, N.; Zaman, M.Z. Effect of modified atmosphere packaging on microbial flora changes in fishery products. *Int. Food Res. J.* **2013**, *20*, 17–26.

20. Cyprian, O.; Lauzon, H.L.; Jóhannsson, R.; Sveinsdóttir, K.; Arason, S.; Martinsdóttir, E. Shelf life of air and modified atmosphere-packaged fresh tilapia (*Oreochromis niloticus*) fillets stored under chilled and superchilled conditions. *Food Sci. Nutr.* **2013**, *1*, 130–140. [CrossRef] [PubMed]

21. Grujić, S.; Grujić, R.; Kovačić, K. Effects of modified atmosphere packaging on quality and safety of fresh meat. *Quality of Life (Banja Luka)* **2010**, *2–4*, 121–133. [CrossRef]

22. Rennie, T.J.; Sunjka, P.S. Modified atmosphere for storage, transportation, and packaging. In *Novel Postharvest Treatments of Fresh Produce*; Pareek, S., Ed.; CRC Press/Taylor and Francis Group: Boca Raton, FL, USA, 2018; pp. 433–480.

23. Boziaris, I.S.; Parlapani, F.F. Specific spoilage organisms (SSOs) in fish. In *The Microbiological Quality of Food*, 1st ed.; Bevilacqua, A., Corbo, M.R., Sinigaglia, M., Eds., Woodhead Publishing Ltd.: Cambridge, UK, 2017; pp. 61–98.

24. Gram, L.; Dalgaard, P. Fish spoilage bacteria – problems and solutions. *Curr. Opin. Biotechnol.* **2002**, *13*, 262–266. [CrossRef]

25. Gram, L.; Huss, H.H. Microbiological spoilage of fish and fish products. *Int. J. Food Microbiol.* **1996**, *33*, 121–127. [CrossRef]

26. Fraser, O.P.; Sumar, S. Compositional changes and spoilage in fish (part II)—Microbiological induced deterioration. *Nutr. Food Sci.* **1998**, *98*, 325–329. [CrossRef]

27. Kuuliala, L.; Abatih, E.; Ioannidis, A.-G.; Vanderroost, M.; De Meulenaer, B.; Ragaert, P.; Devlieghere, F. Multivariate statistical analysis for the identification of potential seafood spoilage indicators. *Food Control* **2018**, *84*, 49–60. [CrossRef]

28. Jouki, M.; Yazdi, F.T.; Mortazavi, S.A.; Koocheki, A.; Khazaei, N. Effect of quince seed mucilage edible films incorporated with oregano or thyme essential oil on shelf-life extension of refrigerated rainbow trout fillets. *Int. J. Food Microbiol.* **2014**, *174*, 88–97. [CrossRef] [PubMed]

29. Kazemi, S.M.; Rezaei, M. Antimicrobial effectiveness of gelatin-alginate film containing oregano essential oil for fish preservation. *J. Food Saf.* **2015**, *35*, 482–490. [CrossRef]

30. Volpe, M.G.; Siano, F.; Paolucci, M.; Sacco, A.; Sorrentino, A.; Malinconico, M.; Varricchio, E. Active edible coating effectiveness in shelf-life enhancement of trout (*Oncorhynchusmykiss*) fillets. *LWT-Food Sci. Technol.* **2015**, *60*, 615–622. [CrossRef]

31. Yıldız, P.O.; Yangılar, F. Effects of different whey protein concentrate coating on selected properties of rainbow trout (*Oncorhynchus mykiss*) during cold storage (4 °C). *Int. J. Food Prop.* **2016**, *19*, 2007–2015. [CrossRef]

32. Yu, D.; Jiang, Q.; Xu, Y.; Xia, W. The shelf life extension of refrigerated grass carp (*Ctenopharyngodon idellus*) fillets by chitosan coating combined with glycerol monolaurate. *Int. J. Biol. Macromol.* **2017**, *101*, 448–454. [CrossRef] [PubMed]

33. Shokri, S.; Ehsani, A. Efficacy of whey protein coating incorporated with lactoperoxidase and α-tocopherol in shelf life extension of Pike-Perch fillets during refrigeration. *LWT-Food Sci. Technol.* **2017**, *85*, 225–231. [CrossRef]

34. Carrión-Granda, X.; Fernández-Pan, I.; Rovira, J.; Maté, J.I. Effect of antimicrobial edible coatings and modified atmosphere packaging on the microbiological quality of cold stored hake (*Merluccius merluccius*) fillets. *J. Food Qual.* **2018**. [CrossRef]

35. Semeniuc, C.A.; Socaciu, M.I.; Socaci, S.A.; Mureşan, V.; Fogarasi, M.; Rotar, A.M. Chemometric comparison and classification of some essential oils extracted from plants belonging to Apiaceae and Lamiaceae families based on their chemical composition and biological activities. *Molecules* **2018**, *23*, 2261. [CrossRef] [PubMed]

36. Chouhan, S.; Sharma, K.; Guleria, S. Antimicrobial activity of some essential oils-present status and future perspectives. *Medicines (Basel)* **2017**, *8*, 5. [CrossRef]

37. Burt, S. Essential oils: Their antibacterial properties and potential applications in foods – a review. *Int. J. Food Microbiol.* **2004**, *94*, 223–253. [CrossRef] [PubMed]

38. Kuorwel, K.K.; Cran, M.J.; Sonneveld, K.; Miltz, J.; Bigger, S.W. Essential oils and their principal constituents as antimicrobial agents for synthetic packaging films. *J. Food Sci.* **2011**, *76*, R164–R177. [CrossRef] [PubMed]

39. Gómez-Estaca, J.; López de Lacey, A.; Gómez-Guillén, M.C.; López-Caballero, M.E.; Monter, P. Antimicrobial activity of composite edible films based on fish gelatin and chitosan incorporated with clove essential oil. *J. Aquat. Food Prod. Technol.* **2009**, *18*, 46–52. [CrossRef]

40. Han, Y.T.; Tammineni, N.; Ünlü, G.; Rasco, B.; Nindo, C. Inhibition of *Listeria monocytogenes* on rainbow trout (*Oncorhynchus mykiss*) using trout skin gelatin edible films containing nisin. *J. Food Chem. Nutr.* **2013**, *1*, 6–15.

41. Min, B.J.; Oh, J.H. Antimicrobial activity of catfish gelatin coating containing origanum (*Thymus capitatus*) oil against gram-negative pathogenic bacteria. *J. Food Sci.* **2009**, *74*, M143–M148. [CrossRef] [PubMed]

42. Ojagh, S.M.; Rezaei, M.; Rzavi, S.H.; Hosseini, S.M.H. Effect of chitosan coatings enriched with cinnamon oil on the quality of refrigerated rainbow trout. *Food Chem.* **2010**, *1*, 193–198. [CrossRef]

43. Li, T.; Li, J.; Hu, W.; Li, X. Quality enhancement in refrigerated red drum (*Sciaenops ocellatus*) fillets using chitosan coatings containing natural preservatives. *Food Chem.* **2013**, *138*, 821–826. [CrossRef] [PubMed]

44. Qiu, X.; Chen, S.; Liu, G.; Yang, Q. Quality enhancement in the Japanese sea bass (*Lateolabrax japonicas*) fillets stored at 4 °C by chitosan coating incorporated with citric acid or licorice extract. *Food Chem.* **2014**, *162*, 156–160. [CrossRef] [PubMed]

45. Bahram, S.; Rezaie, M.; Soltani, M.; Kamali, A.; Abdollahi, M.; Ahmadabad, M.K.; Nemati, M. Effect of whey protein concentrate coating cinamon oil on quality and shelf life of refrigerated Beluga Sturegeon (*Huso huso*). *J. Food Qual.* **2016**, *39*, 743–749. [CrossRef]

46. Souza, B.W.S.; Cerqueira, M.A.; Ruiz, H.A.; Martins, J.T.; Casariego, A.; Teixeira, J.A.; Vicente, A.A. Effect of chitosan-based coatings on the shelf life of salmon (*Salmo salar*). *J. Agric. Food Chem.* **2010**, *58*, 11456–11462. [CrossRef] [PubMed]

47. Feng, X.; Bansal, N.; Yang, H. Fish gelatin combined with chitosan coating inhibits myofibril degradation of golden pomfret (*Trachinotus blochii*) fillet during cold storage. *Food Chem.* **2016**, *200*, 283–292. [CrossRef] [PubMed]

48. Vásconez, M.B.; Flores, S.K.; Campos, C.A.; Alvarado, J.; Gerschenson, L.N. Antimicrobial activity and physical properties of chitosan–tapioca starch based edible films and coatings. *Food Res. Int.* **2009**, *42*, 762–769. [CrossRef]

49. Ramezani, A.; Zarei, M.; Raminnejad, N. Comparing the effectiveness of chitosan and nanochitosan coatings on the quality of refrigerated silver carp fillets. *Food Control* **2015**, *51*, 43–48. [CrossRef]

50. Cardoso, L.G.; Santos, J.C.P.; Camilloto, G.P.; Miranda, A.L.; Druzian, J.I.; Guimarães, A.G. Development of active films poly (butylene adipate co-terephthalate)—PBAT incorporated with oregano essential oil and application in fish fillet preservation. *Ind. Crop. Prod.* **2017**, *108*, 388–397. [CrossRef]

51. Rollini, M.; Nielsen, T.; Musatti, A.; Limbo, S.; Piergiovanni, L.; Munoz, P.H.; Gavara, R. Antimicrobial performance of two different packaging materials on the microbiological quality of fresh salmon. *Coatings* **2016**, *6*, 6. [CrossRef]

52. Semeniuc, C.A.; Mandrioli, M.; Rodriguez-Estrada, M.T.; Muste, S.; Lercker, G. Thiobarbituric acid reactive substances in flavoured phytosterol-enriched drinking yogurts during storage: formation and matrix interferences. *Eur. Food Res. Tech.* **2016**, *242*, 431–439. [CrossRef]

53. Castro, P.; Millán, R.; Penedo, J.C.; Sanjuán, E.; Santana, A.; Caballero, M.J. Effect of storage conditions on total volatile base nitrogen determinations in fish muscle extracts. *J. Aquat. Food Prod. Technol.* **2012**, *21*, 519–523. [CrossRef]

54. Sikorski, Z.E.; Kołakowska, A.; Pan, B.S. The nutritive composition of the major groups of marine food organisms. In *Seafood: Resources, Nutritional Composition, and Preservation*; Sikorski, Z.E., Ed.; CRC Press: Boca Raton, FL, USA, 1990; pp. 55–76.

55. Ruiz-Capillas, C.; Herrero, A.M.; Jiménez-Colmenero, F. Determination of volatile nitrogenous compounds: ammonia, total volatile basic nitrogen, and trimethylamine. In *Flow Injection Analysis of Food Additives*; Ruiz-Capillas, C., Nollet, L.M.L., Eds.; CRC Press/Taylor and Francis Group: Boca Raton, FL, USA, 2015; pp. 659–674.

56. European Union. Commission Regulation (EC) No. 2074/2005 of 5 December 2005 laying down implementing measures for certain products under Regulation (EC) No. 853/2004 of the European Parliament and of the Council and for the organisation of official controls under Regulation (EC) No. 854/2004 of the European Parliament and of the Council and Regulation (EC) No. 882/2004 of the European Parliament and of the Council, derogating from Regulation (EC) No. 852/2004 of the European Parliament and of the Council and amending Regulations (EC) No. 853/2004 and (EC) No. 854/2004. *Off. J. Eur. Union* **2005**, *48*, 27–59.

57. *Non-Sensory Assessment of Fish Quality*; Torry Advisory Note No. 92; Torry Research Station, MAFF: Aberdeen, UK, 1989; pp. 1–6.

58. Bystedt, J.; Swenne, L.; Aas, H.W. Determination of trimethylamine oxide in fish muscle. *J. Sci. Food Agric.* **1959**, *10*, 301–304. [CrossRef]

59. Cheng, J.H.; Sun, D.W.; Pu, H.; Zhu, Z. Development of hyperspectral imaging coupled with chemometric analysis to monitor *K* value for evaluation of chemical spoilage in fish fillets. *Food Chem.* **2015**, *185*, 245–253. [CrossRef] [PubMed]

60. European Union. Regulation (EC) No. 1935/2004 of the European Parliament and of the Council of 27 October 2004 on materials and articles intended to come into contact with food and repealing directives 80/590/EEC and 89/109/EEC. *Off. J. Eur. Union* **2004**, *47*, 4–17.

61. European Union. Commission Regulation (EC) No. 450/2009 of 29 May 2009 on active and intelligent materials and articles intended to come into contact with food. *Off. J. Eur. Union* **2009**, *52*, 3–11.

62. European Union. Regulation (EC) No. 1333/2008 of the European Parliament and of the Council of 16 December 2008 on food additives. *Off. J. Eur. Union* **2008**, *51*, 16–33.

Article

Architectured Cu–TNTZ Bilayered Coatings Showing Bacterial Inactivation under Indoor Light and Controllable Copper Release: Effect of the Microstructure on Copper Diffusion

Akram Alhussein [1,*], **Sofiane Achache** [1], **Regis Deturche** [2] **and Sami Rtimi** [3,*]

[1] ICD-LASMIS, Université de Technologie de Troyes, Antenne de Nogent, Pôle Technologique Sud Champagne, 52800 Nogent, France; sofiane.achache@utt.fr
[2] ICD-L2N, Université de Technologie de Troyes, 12 Rue Marie Curie, 10010 Troyes, France; regis.deturche@utt.fr
[3] Ecole Polytechnique Fédérale de Lausanne, EPFL-STI-IMX-LTP, 1015 Lausanne, Switzerland
*   Correspondence: akram.alhussein@utt.fr (A.A.); rtimi.sami@gmail.com (S.R.); Tel.: +33-35-1591174 (A.A.); +41-21-6936150 (S.R.); Fax: +33-32-5717676 (A.A.)

Received: 16 May 2020; Accepted: 15 June 2020; Published: 19 June 2020

Abstract: A Ti–23Nb–0.7Ta–2Zr–1.2O alloy (at %), called "gum metal", was deposited by direct-current magnetron sputtering (DCMS) on an under layer of copper. By varying the working pressure during the deposition, columnar TNTZ (Ti–Nb–Ta–Zr) nanoarchitectures were obtained. At low working pressures, the upper layer was dense with a coarse surface (R_a = 12 nm) with a maximum height of 163 nm; however, the other samples prepared at high working pressures showed columnar architectures with voids and an average roughness of 4 nm. The prepared coatings were characterized using atomic force microscopy (AFM) for surface topography, energy dispersive X-ray spectroscopy (EDX) for atomic mapping, scanning electron microscopy (SEM) for cross-section imaging, contact angle measurements for hydrophilic/hydrophobic balance of the prepared surfaces, and X-ray diffraction (XRD) for the crystallographic structures of the prepared coatings. The morphology and the density of the prepared coatings were seen to influence the hydrophilic properties of the surface. The antibacterial activity of the prepared coatings was tested in the dark and under low-intensity indoor light. Bacterial inactivation was seen to happen in the dark from samples presenting columnar nanoarchitectures. This was attributed to the diffusion of copper ions from the under layer. To verify the copper release from the prepared samples, an inductively coupled plasma mass spectrometer (ICP-MS) was used. Additionally, the atomic depth profiling of the elements was carried out by X-ray photoelectron spectroscopy (XPS) for the as-prepared samples and for the samples used for bacterial inactivation. The low amount of copper in the bulk of the TNTZ upper layer justifies its diffusion to the surface. Recycling of the antibacterial activity was also investigated and revealed a stable activity over cycles.

Keywords: titanium-based thin films; copper; magnetron sputtering; super-elastic coatings; *E. coli* inactivation

---

1. Introduction

Thin films of titanium alloys are highly used in biomedical fields due to their lightness, low elasticity modulus, good mechanical strength, biocompatibility, and anticorrosion behavior [1,2]. These properties depend on the film elemental composition [3]. The favorable phase of β-type Ti-based thin films presents a body-centered cubic (bcc) structure, low density, and Young's modulus. β-type Ti-based thin films containing nontoxic elements and non-allergic properties have been largely developed during these last years. The enhancement of their mechanical properties and super-elastic performance is one of the most important issues [2–5]. After developing Ti–Nb–Zr–Ta (TNTZ) bulk alloys, in particular the multifunctional Ti–23Nb–0.7Ta–2Zr–1.2O alloy (at %), called "gum metal", it has been shown that these alloys can be used in the form of super-elastic thin films. These films present an important potential for medical applications and/or microactuation, such as for stents for neurovascular blood vessels or membrane-based micropumps [6–9].

In the present work, two superposed layers, namely gum metal (GM) and copper thin films, were deposited on glass substrates by direct-current magnetron sputtering (DCMS) of an alloying target Ti–23Nb–0.7Ta–2Zr (at %) and a pure Cu target. In previous studies [10–12], we reported the beneficial effect of Cu ions on enhancing the catalytic/photocatalytic bacteria inactivation when added to a gum metal thin film (GMTF). Many copper-based coatings such as TiO_2–Cu, Cu, Ti, Nb, and their oxides were developed and investigated, mainly for their antibacterial properties. In our previous studies [13–15], we reported the mechanical properties of a superelastic Ti-based GMTF showing increased bacterial inactivation under light concomitant to a biocompatible property in darkness [16,17]. We also showed the effect of copper content on the structure, morphology, and mechanical properties of GMTF [18]. We also showed that the TNTZ–Cu 8.3 at % Cu sample with the highest roughness (RMS value of 20.1 nm) led to the highest bactericidal activity.

The present study focuses on the bacterial inactivation of sputtered GMTF films deposited with different pressure values on copper diffusion to the surface mediating the bacterial inactivation. The application of different pressure values in this study was designed to lead to different layer densities and microarchitectures (disordered tubes or arranged columns). The influence of the morphology of the TNTZ upper layer allowing copper diffusion is investigated in the dark and under indoor visible-light irradiation. The interfacial interaction, potential, and pH between the bacterial cells on the sputtered films are reported within the disinfection time on these innovative bilayered Cu–TNTZ coatings. The sustainable use (reusability) of the sputtered samples was also studied, showing a long operational lifetime of the gradual release of copper from the sputtered films. The reported thin films show a potential application in biomedical settings, leading to self-sterilization of surfaces without harmful side-effects related to the excess of toxic ions, especially in circumstances of increased infection risks in hospitals. The novelty of this work resides in the preparation of a coating allowing fast bacterial inactivation with reduced ion release.

2. Experimental Details

2.1. Deposition Procedures of the Nanoarchitected Films of Cu and TNTZ

Bilayer films composed of Cu and gum metal (TNTZ) were consecutively deposited on glass substrates and silicon wafers (1 0 0) at floating temperature by DCMS from two targets (99.9% purity): pure Cu and Ti–23Nb–0.7Ta–2Zr (at %), respectively. Two similar magnetrons joined to rectangular targets $200 \times 100 \times 6$ mm^3 in size were used. The chamber was evacuated down to a pressure of 4×10^{-4} Pa. The target-to-sample distance was kept constant at 8 cm. After initial optimization, the first layer of copper was sputtered for 20 s and then the second layer of TNTZ was carried out for 40 s. The thickness of the Cu–TNTZ bilayer films was approximately 1 µm. The sample holder rotation speed was set to 12 rpm. The working pressure was varied between 0.15 and 2 Pa, corresponding to the argon flow rate of 15–140 sccm. Figure 1 illustrates the setup used for the deposition, and Table 1 illustrates the deposition parameters of Cu–TNTZ bilayered thin films.

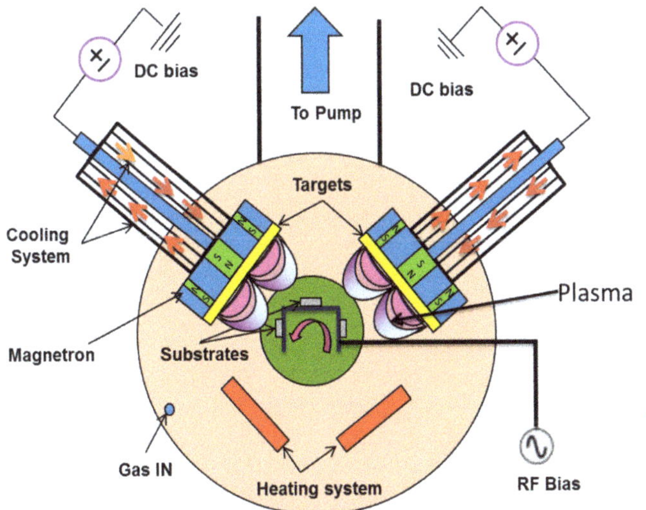

Figure 1. Schematic view of the setup used for the bilayered film deposition.

Table 1. Elaboration conditions of Cu–TNTZ (Ti–Nb–Ta–Zr) bilayered thin films.

Sample	Cu Under Layer		TNTZ Upper Layer	
	Argon Flow Rate (sccm)	Working Pressure (Pa)	Argon Flow Rate (sccm)	Working Pressure (Pa)
I			15	0.15
II	15	0.15	50	0.5
III			90	1
IV			140	2

2.2. Physical and Chemical Characterization of the Sputtered Samples by Energy Dispersive X-Ray Spectroscopy (EDX), SEM, Atomic Force Microscopy (AFM), and XRD

The elemental compositions of the TNTZ films were obtained by energy dispersive X-ray spectroscopy (EDX) using Quantax Esprit software version 2.0 (Bruker Nano GmbH, Berlin, Germany). The bilayer Cu–TNZT cross-section images were observed by a Hitachi SU8030 scanning electron microscope (SEM) (Tokyo, Japan). For each sample, the measurements were carried out on multiple zones of the sample and showed a significant homogeneity. The elements spectra were acquired from SEM images applying an accelerating voltage of 10 kV and a pressure vacuum of 10^{-4} Pa. The surface roughness and the topography of the thin films were obtained by atomic force microscopy (Agilent Technologies 5100 AFM, Stevens Creek, CA, USA). The AFM 3D-images were acquired in noncontact mode at the resonance frequency of the ultrasharp silicon tip and with the set point fixed at 10% of the maximum amplitude. The AFM images of 256 × 256 pixels were obtained with a scanning speed of 0.4 s/line (acquisition of 256 points/line).

The microstructure of the thin films was analyzed by X-ray diffraction (XRD) using a Bruker D8 Advance diffractometer (CuKα radiation line, Karlsruhe, Germany). The hydrophilic/hydrophobic properties of the sputtered films were evaluated by contact angle measurements based on the static sessile drop method. Water droplets (3 mL) were dropped on the films and the contact angles were measured. The droplet profiles were acquired using a camera (SCA software, version 4.5.14) for OCA and PCA Drpo6) aligned with the sample and a backlighting source. For this study, Si wafers were used in order to observe the film morphology and to quantify the film elemental composition. Glass substrates (Leica, Germany) were used to perform the other characterizations and antibacterial tests.

2.3. Bacterial Adhesion to the Coating Surface, Bacterial Inactivation Testing, and the Sustainability of the Coatings

Escherichia coli (*E. coli* K12) used in this study was obtained from DSMZ (Braunschweig, Germany). The bacterial adhesion to the TNTZ and Cu–TNTZ was carried out by immersing the samples into 5 mL *E. coli* suspension (initial concentration of 2.2×10^6 CFU/mL). The tube was then horizontally shacked gently for 4 h in an incubator at 37 °C in the dark [19,20]. The non-adhered and the weakly adhered bacteria to the surface were washed out with a phosphate buffer solution (pH 7.2). The attached cells were then detached from the surface using ultrasonication (50 W, CA, USA) for 15 min, and the number of viable cells was determined by plate-count agar.

For the bacteria inactivation testing, aliquots of 100 μL (initial concentration of 4.2×10^6 CFU/mL) in saline solution (NaCl/KCl, 8/0.8 g/L at pH 7.2) were placed on the samples that were placed in glass petri dishes at room temperature. Then, the samples were transferred into a sterile tube containing saline solution and subsequently mixed thoroughly using a Vortex (Staufen, Germany) for 2 min. Serial dilutions were made. A sample of 100 μL from each dilution was pipetted onto a nutrient agar plate (Zurich, Switzerland) and spread over the surface of the plate. Agar plates were incubated lid down for 24 h at 37 °C before counting. Three independent assays were done for each sample. The statistical analysis of the results was performed for the CFU values calculating the standard deviation ($n = 5\%$). Irradiation of the samples was carried out under tubular Osram Lumilux (Munich, Germany) 18W/827 light (4.2 mW/cm^2) emitting in the range of 360–700 nm.

Bacterial inactivation recycling was carried out under light and in the dark. After each cycle, the sample was washed three times with distilled water and dried between each washing. All the bacterial inactivation tests were carried out in triplicate.

2.4. Copper Release and Atomic Etching of the Architectured Samples

A Finnigan^{TM} inductively coupled plasma mass spectrometer (ICP-MS) (Zug, Switzerland) was used to detect the Cu released during the bacterial loss of viability runs. In addition, to test the long-term stability, the coated specimens were immersed in 20 ml PBS and kept at 37 °C for 1, 2, 3, and 5 days. This instrument was equipped with a double-focusing reverse geometry mass spectrometer having an extremely low background signal and high ion-transmission coefficient with a detection limit of 0.2 ppb. Each specimen was tested in triplicate.

X-ray photoelectron spectroscopy (XPS) analysis of the films was determined using an AXIS Nova photoelectron spectrometer (Kratos Analytical, Manchester, UK) provided with a monochromatic AlKα ($hv = 1486.6$ eV) anode operated with a voltage of 15 kV and a power of 400 W. The carbon C1s line position at 284.6 eV was used as the reference to correct for charging effects. The base pressure below 5×10^{-7} Pa was maintained during the measurements. No argon sputtering was considered to clean the specimen surfaces. The surface atomic concentration percentage for each element was determined from peak areas using the known sensitivity factors for each element on the coated surfaces. The etching of the film surfaces was carried out by Ar ions of 5 kV reaching a depth of ~50 nm (~250 layers).

3. Results and Discussion

3.1. Elemental Composition, Morphology, and Topography of Cu–TNTZ Thin Films Deposited under Different Pressures

The elemental composition of the TNTZ films deposited on a copper layer was determined by EDX. The films were chemically homogeneous and composed similarly of the TNTZ target's elements: Ti–23Nb–0.7Ta–2Zr (at %). Figure 2 shows fractured cross-section SEM images of different Cu–TNTZ bilayered samples sputtered under different working pressures.

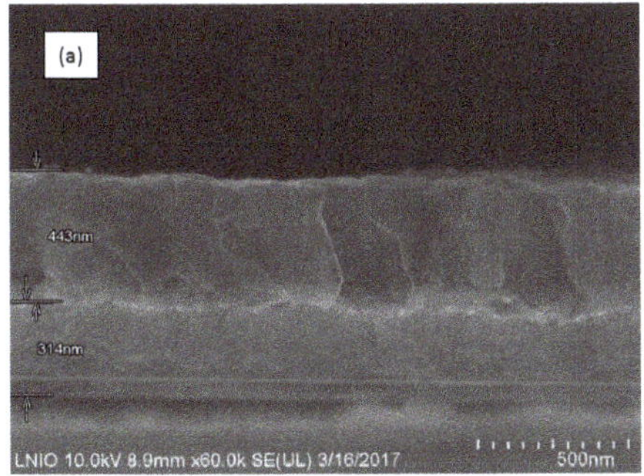
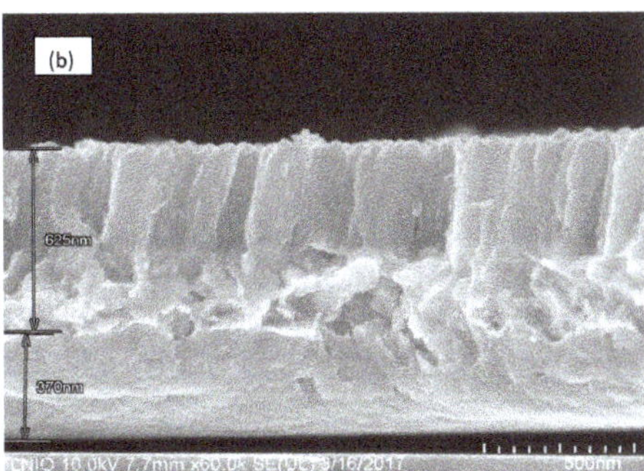
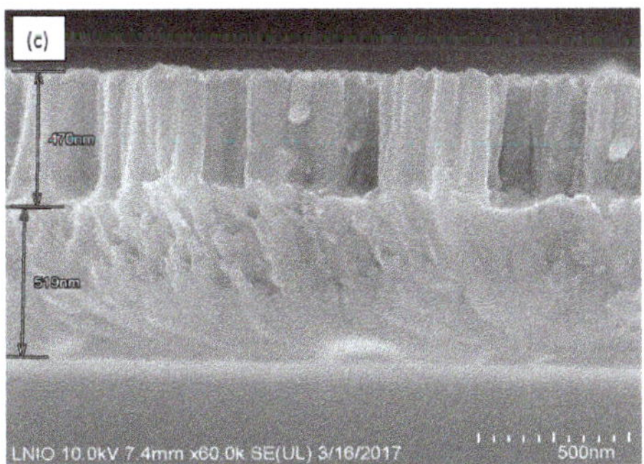

Figure 2. Cross-section SEM images of architectured Cu–TNTZ films with different morphologies sputtered at different pressures: (**a**) I—0.15 Pa, (**b**) II—0.50 Pa, and (**c**) III—1 Pa.

A dense layer of Cu was first deposited under a low constant pressure of 0.15 Pa, then an upper layer of TNTZ was sputtered under different deposition pressures. It is readily seen that by increasing the deposition pressure from 0.15 to 2 Pa, the morphology of the TNTZ films changes. Under low pressures, dense TNTZ upper films were obtained. However, when increasing the deposition pressure, columnar structures were achieved. This structural shift was attributed to the loss of adatom energy and surface mobility because of the collision with argon ions in the plasma phase.

Figure 3 shows the topography of the samples using atomic force microscopy (AFM). The 3D and 2D surface mappings of the TNTZ films (Figure 3) show that the growth of films forms hills and valleys. Except for Sample I, the roughness average of all other Cu–TNTZ bilayer samples is 5 nm. Sample I has the coarsest surface (R_a = 12 nm) with a maximum height of 155 nm.

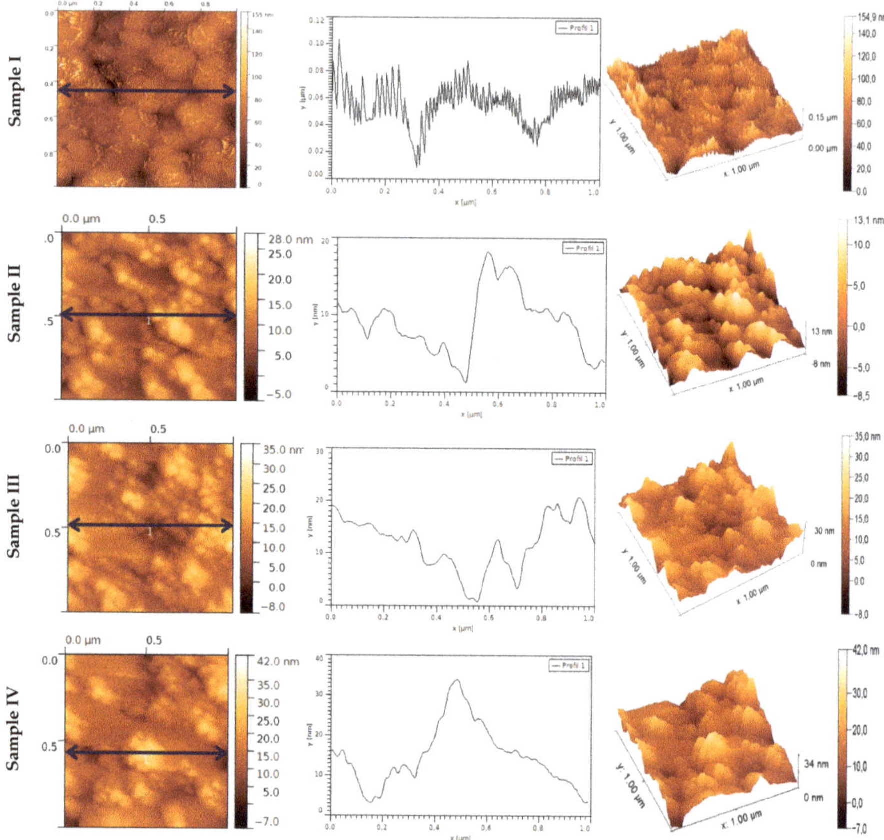

Figure 3. 2D and 3D atomic force microscopy (AFM) of TNTZ thin films sputtered under different pressure conditions. For more details, please see Table 1.

3.2. Crystalline Structure and Hydrophobic/Hydrophilic Properties of the TNTZ Films

Figure 4 shows the XRD patterns of the TNTZ films deposited on a copper under layer at different deposition pressures. It is readily seen from Figure 4 that the TNTZ films present a β-metastable phase. Similar results were observed in our previous study [13]. Furthermore, the crystallinity of TNTZ films decreases with the increase of deposition pressure. The TNTZ film deposited at 2 Pa shows a low-intensity single peak, while others deposited at lower pressures exhibit sharper peaks. Similar trends were observed in our previous study [13]. This can be attributed to the decrease of a mean free path leading to the lower adatom surface mobility. Thus, the adatoms cannot migrate for a larger distance, promoting the formation of fine grains.

The presence of peaks related to the Cu underlayer is due to the interaction of X-rays with both layers of Cu and TNTZ. The penetration of X-rays within the material is estimated to be ~1 µm; thus, crystalline phases of both layers show up on the diffractogram [21]. The increase of copper peak intensity with the decrease of deposition pressure could be explained by the decrease of the thickness of the upper TNTZ layer. It is obvious that the decrease of deposition pressure in a sputtering process leads to a decrease of film thickness. Therefore, the diffraction peaks attributed to the copper layer are more remarkable at lower deposition pressures.

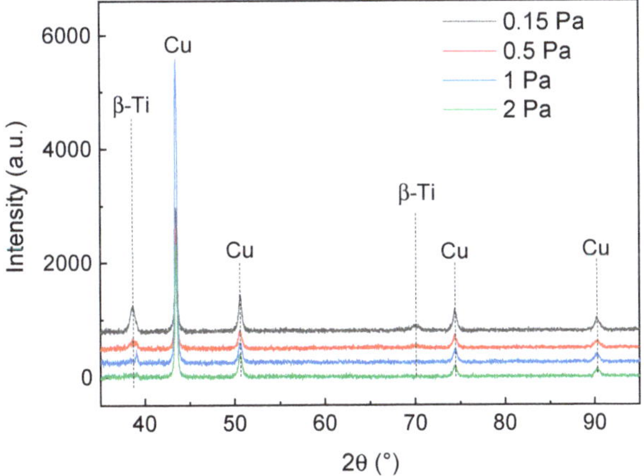

Figure 4. X-ray diffraction (XRD) of architectured Cu–TNTZ samples sputtered at different pressures. JCPDS 00-004-0836 and JCPDS 00-044-1288 were used for Cu and β-Ti, respectively.

The hydrophilic/hydrophobic properties of the architectured Cu–TNTZ films were evaluated via the contact angles between the films deposited on glass substrates and the water droplets. Water droplets of 3 μL in volume fallen on the film surfaces were measured, as shown in Figure 5. The contact angles varied between 81° and 102°. The roughness, the morphology, and the density of film (as previously shown in Figures 2 and 3) influenced the hydrophilic properties of the surface. The denser and coarser the film was (Sample I, Figures 2 and 3), the lower the contact angle was (81°). For the other films deposited under higher pressures (in the range of 1–2 Pa), the samples exhibited contact angles of 100°–102°. As is shown in Figures 2 and 3, these latter films showed columnar architectures and low roughness. The film prepared at 0.5 Pa showed a contact angle value of 92°.

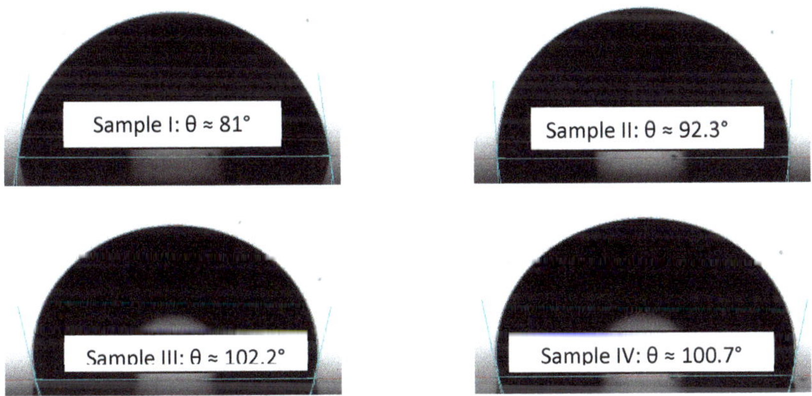

Figure 5. Contact angles between drops of water and Cu–TNTZ bilayer films.

3.3. Bacterial Adhesion and Inactivation, and the Atomic Depth Profiling before and after Bacterial Inactivation

Bacterial adhesion at the interface of the TNTZ-sputtered films with different architectures was investigated. The TNTZ films sputtered at the different conditions described in Table 1, without the copper underlayer, were used to investigate the influence of the interface topography/microstructure on the *E. coli* adhesion. The TNTZ-presenting dense/coarse architecture as shown in Supplementary

Figure S1a exhibited low bacterial adhesion (1.4 $\log_{10}$ CFU) compared to samples showing columnar-shaped microstructures (Figure S1b,c), allowing 2.2 to 2.6 $\log_{10}$ CFU adhesion in the dark. The difference between the samples presenting columnar architectures was not significant in the triplicate experiment carried out. This can be attributed to the hills and valleys at the interface, allowing a larger contact surface with bacteria [22]. Figure 6 shows the possible bacterial interaction with the two different microstructures of sputtered TNTZ thin films. The dense surface presents low contact points with bacteria, leading to the low adhesion ability compared to the columnar counterparts. It is worth mentioning that the chemistry of the surface and its charge may strongly affect the ability of a thin film to adhere to bacteria. Recently, bacterial adhesion to biotic and abiotic surfaces has been investigated in detail [23].

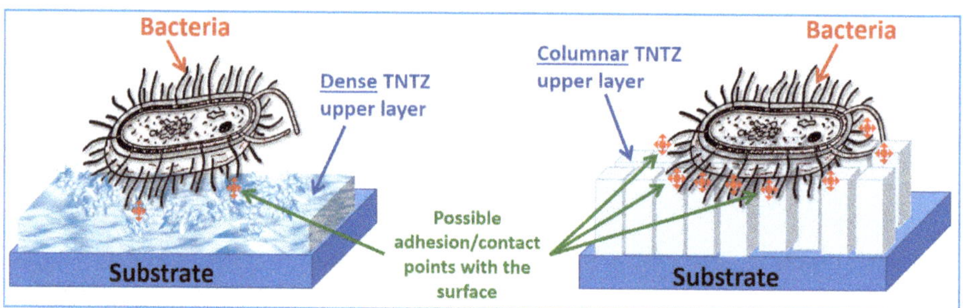

Figure 6. Simplified understanding of the bacterial adhesion to nanoarchitectured surfaces.

Under light, TNTZ films did not exhibit higher bacterial inactivation values compared to their performance in the dark (bacterial adhesion). Under light, dense TNTZ showed 1 log reduction against 1.8 log reduction for columnar TNTZ. This may be due to the oxidation of the forming species at the interface with bacteria forming metal oxides. Under light, these latter oxides generate reactive oxygen species able to inactivate bacteria at their interface. Detailed investigation of this aspect should be carried out to better understand the mechanism of bacterial inactivation under low-intensity indoor light.

Figure 7 shows the bacterial inactivation kinetics on Samples I, II, and III under low-intensity indoor light and in the dark. *E. coli* inactivation (6 log reduction) was seen to happen within 60 min on Samples II and III. Sample I showed 3 log bacterial reduction within 300 min under light (4.2 mW/cm^2).

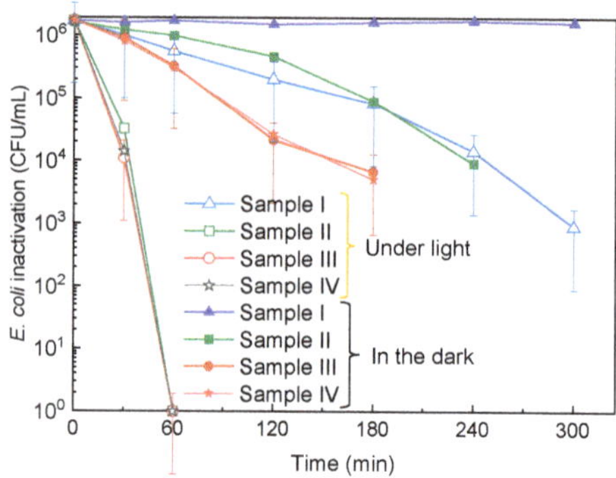

Figure 7. *E. coli* inactivation on bilayered samples under indoor light and in the dark.

This justifies the semiconductor behavior at the interface of Samples I, II, III, and IV, and the accelerated bacterial inactivation under light is photocatalytic. In the dark, Samples II, III, and

IV showed 4 log bacterial reduction within 180–240 min. This suggests the existence of toxic ions on the surfaces leading to bacterial inactivation in the dark. Sample I shows that the compact upper layer did not show bacterial reduction in the dark. Therefore, the columnar nanoarchitectures of Samples II, III, and IV allowed the diffusion of copper ions reaching the surface and enhanced the bacterial inactivation in the dark.

Surface atomic percentages of elements as a function of the depth of the sputtered film were also investigated. As Samples II and III showed the fastest bacterial inactivation, we chose to carry the depth profiling on Sample II. However, it is worth mentioning that Samples III and IV exhibited higher copper content on their surfaces than Sample II. Copper ions were not detected at the surface of Sample I, which means that the copper diffusion was blocked by the dense TNTZ upper layer. Figure 8 shows the depth profiling of Sample II. Copper ions encountered at the coating interface were seen to enhance the bacterial inactivation under light and, more importantly, in the dark. It has been recently reported that sputtered films with Cu in ppm concentrations led to an acceleration of the bacterial inactivation [24].

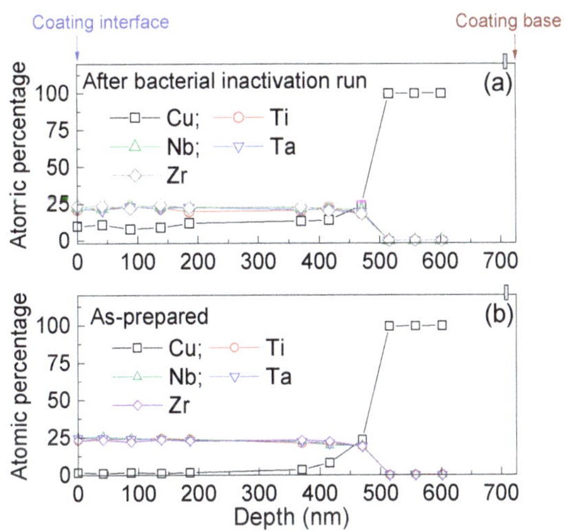

Figure 8. Depth profiling atom distribution in the sputtered layers of Sample II: evidence for copper diffusion in: (**a**) sample used for bacterial inactivation with respect to (**b**) an as-prepared sample.

3.4. Recycling and Ion Release during Bacterial Inaction

Bacterial inactivation recycling was carried out to test the stability of the prepared antibacterial coatings. Figure 9 shows the recycling of Samples II and III under light and in the dark. It is readily seen from Figure 9 that Samples II and III did not show a significant loss of their activities in up to five washing cycles. Coatings prepared by magnetron sputtering present high adhesivity, uniform distribution on the substrate, and controllable nanoarchitecture [25–27]. This allows their practical application in the biomedical field, environmental remediation, and smart surfaces [28,29].

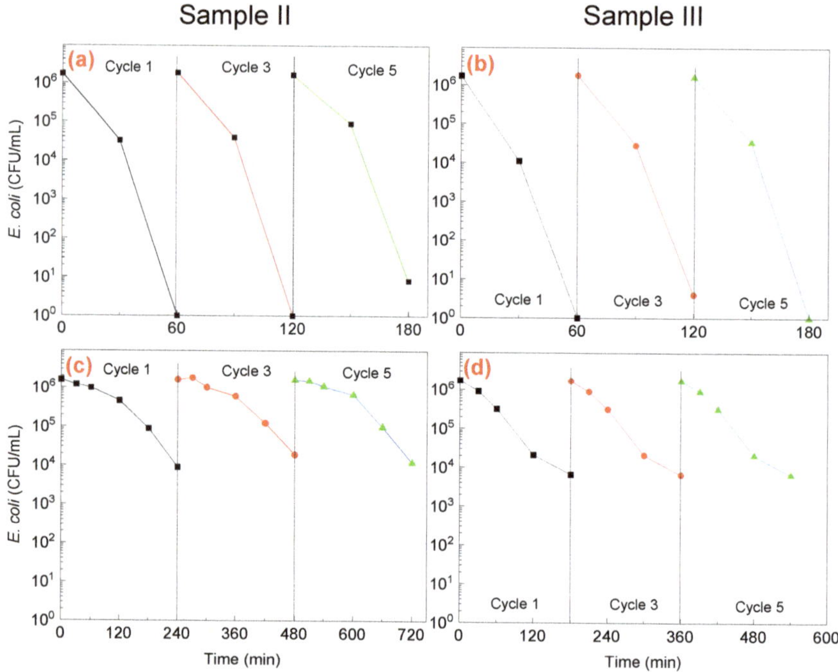

Figure 9. Recycling of *E. coli* inactivation under light and in the dark on Samples II and III: (**a**) Sample II under light, (**b**) Sample III under light, (**c**) Sample II in the dark, and (**d**) Sample III in the dark.

The leached copper ions were determined for the as-prepared samples gently washed with water and after the bacterial inactivation cycles. Figure 10 shows the copper ion release from Samples I, II, III, and IV. Sample IV showed the highest copper release. This can be due to the spaced nanocolumns in its microstructure, allowing higher copper amounts to reach the interface with bacteria. This aspect needs deeper investigation. The leached amounts of copper are far below the toxicity threshold. The nanoarchitecture of the prepared coatings allowed the controllable diffusion of copper ions to the interface, with bacteria leading to their inactivation. Detailed Cu-toxicity studies from these coatings are needed when considering the practical application of these films.

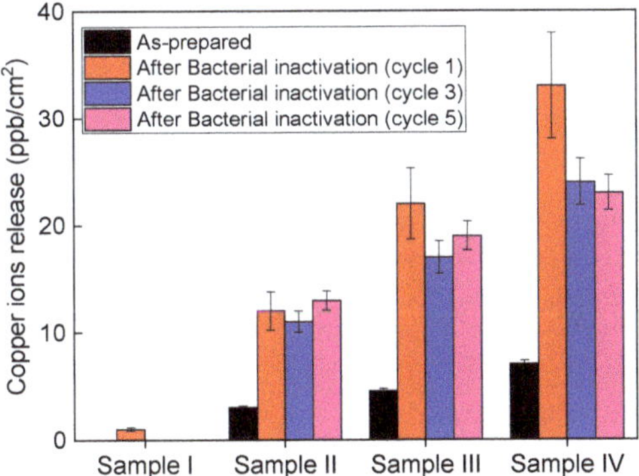

Figure 10. Copper ion release before and after the bacterial inactivation on the different Cu–TNTZ sputtered films.

In our previous study, we showed that TNTZ by itself presents a moderate bacterial inactivation activity [18]. This was attributed to the oxidation of the TNTZ species at the interface, forming oxides able to absorb light and produce reactive oxygen species. Then, we studied the beneficial effect of cosputtering TNTZ with copper/copper oxides, leading to fast bacterial inactivation. This preparation showed fast bacterial inactivation due to the copper species at the interface with bacteria. The preparation led to uncontrollable copper release. In the present study, the copper release seems to be stable over the bacterial inactivation cycles, which reduce the risks of copper toxicity. Figure 11 illustrates the possible mechanism behind the catalytic/photocatalytic activities at the interface of the sputtered nanostructured coatings, especially Sample II, leading to the most controllable copper release.

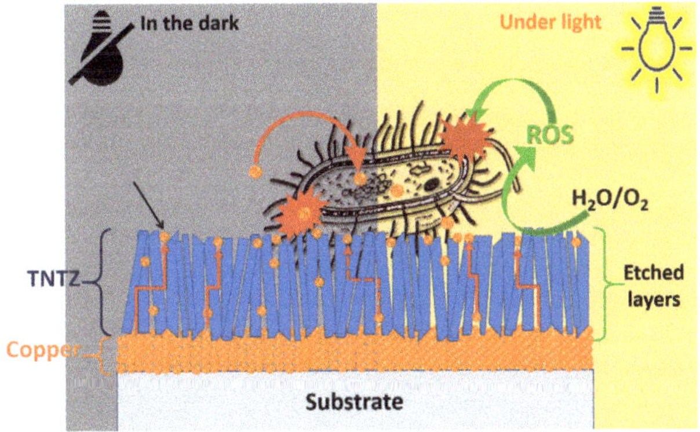

Figure 11. Schematic antibacterial activity at the interface of nanoarchitectured Cu–TNTZ thin films prepared by magnetron sputtering.

4. Conclusions

Super-elastic titanium based TNTZ thin films were deposited by DC magnetron sputtering (DCMS) under different working pressures in the range of 0.15–2 Pa. An intermediate dense layer of pure copper was first deposited at a constant pressure of 0.15 Pa. The morphology, microstructure, and surface topography of these samples were studied. The working pressure in the deposition chamber influenced the morphology of the films, which were dense at 0.15 Pa (low pressure), and columnar architectures were achieved as the pressure values varied from 0.5 to 2 Pa. Bacterial inactivation was seen to happen within 60 min under light for Samples II and III, showing copper ion release. The XPS depth profiling of the sputtered bilayers revealed the existence of some copper species in the bulk of the TNTZ upper layer, justifying its diffusion to the surface. Recycling was seen to happen with a stable fashion. The prepared coatings present potential application for indoor environmental prevention, smart surfaces, and biomedical tools (screens, incubators, and other glass-made surfaces able to harbor pathogens).

Supplementary Materials: The following are available online at http://www.mdpi.com/2079-6412/10/6/574/s1, Figure S1: SEM images of TNTZ coatings sputtered at different working pressures.

Author Contributions: Conceptualization, A.A. and S.R.; methodology, A.A., S.A., R.D., and S.R.; validation, A.A. and S.R.; formal analysis, A.A. and S.R.; investigation, A.A., S.A., R.D., and S.R.; data curation, A.A. and S.R.; writing—original draft preparation, A.A. and S.R.; writing—review and editing, A.A. and S.R. All authors discussed the results and implications and commented on the manuscript at all stages. All authors have read and agreed to the published version of the manuscript.

Funding: This research received no external funding.

Acknowledgments: S. Rtimi thanks H. Hofmann (IMX-EPFL) for the samples' characterization.

Conflicts of Interest: The authors declare no conflict of interest.

References

1. Kim, H.Y.; Sasaki, T.; Okutsu, K.; Kim, J.I.; Inamura, T.; Hosoda, H.; Miyazaki, S. Texture and shape memory behavior of Ti–22Nb–6Ta alloy. *Acta Mater.* **2006**, *54*, 423–433. [CrossRef]

2. Tane, M.; Akita, S.; Nakano, T.; Hagihara, K.; Umakoshi, Y.; Niinomi, M.; Nakajima, H. Peculiar elastic behavior of Ti–Nb–Ta–Zr single crystals. *Acta Mater.* **2008**, *56*, 2856–2863. [CrossRef]

3. Hu, Q.M.; Li, S.J.; Hao, Y.L.; Yang, R.; Johansson, B.; Vitos, L. Phase stability and elastic modulus of Ti alloys containing Nb, Zr, and/or Sn from first-principles calculations. *Appl. Phys. Lett.* **2008**, *93*, 121902. [CrossRef]

4. Li, S.J.; Cui, T.C.; Hao, Y.L.; Yang, R. Fatigue properties of a metastable β-type titanium alloy with reversible phase transformation. *Acta Biomater.* **2008**, *4*, 305–317. [CrossRef]

5. Sikka, S.K.; Vohra, Y.K.; Chidambaram, R. Omega phase in materials. *Prog. Mater. Sci.* **1982**, *27*, 245–310. [CrossRef]

6. Niinomi, M. Fatigue performance and cyto-toxicity of low rigidity titanium alloy, Ti–29Nb–13Ta–4.6 Zr. *Biomaterials* **2003**, *24*, 2673–2683. [CrossRef]

7. Laheurte, P.; Prima, F.; Eberhardt, A.; Gloriant, T.; Wary, M.; Patoor, E. Mechanical properties of low modulus β titanium alloys designed from the electronic approach. *J. Mech. Behav. Biomed. Mater.* **2010**, *3*, 565–573. [CrossRef]

8. Saito, T.; Furuta, T.; Hwang, J.H.; Kuramoto, S.; Nishino, K.; Suzuki, N. Multifunctional alloys obtained via a dislocation-free plastic deformation mechanism. *Science* **2003**, *300*, 464. [CrossRef]

9. Otsuka, K.; Ren, X. Recent developments in the research of shape memory alloys. *Intermetallics* **1999**, *7*, 511. [CrossRef]

10. Rtimi, S.; Giannakis, S.; Sanjines, R.; Pulgarin, C.; Bensimon, M.; Kiwi, J. Insight on the photocatalytic bacterial inactivation by co-sputtered TiO_2–Cu in aerobic and anaerobic conditions. *Appl. Catal. B Environ.* **2016**, *182*, 277–285. [CrossRef]

11. Baghriche, O.; Rtimi, S.; Pulgarin, C.; Sanjines, R.; Kiwi, J. Effect of the spectral properties of TiO_2, Cu, TiO_2/Cu sputtered films on the bacterial inactivation under low intensity actinic light. *J. Photochem. Photobiol. A Chem.* **2013**, *251*, 50–56. [CrossRef]

12. Wojcieszak, D.; Mazur, M.; Kaczmarek, D.; Szponar, B.; Grobelny, M.; Kalisz, M.; Pelczarska, A.; Szczygiel, I.; Poniedzialek, A.; Osekowska, M. Structural and surface properties of semitransparent and antibacterial (Cu, Ti, Nb) Ox coating. *Appl. Surf. Sci.* **2016**, *380*, 159–164. [CrossRef]

13. Brown, R.D.; Boggs, J.E.; Hilderbrandt, R.; Lim, K.; Mills, I.M.; Niktin, E.; Palmer, M.H. Gum Metal thin films obtained by magnetron sputtering of a Ti-Nb-Zr-Ta target. *Mater. Sci. Eng. A* **2016**, *673*, 492–502.

14. Achache, S.; Lamri, S.; Arab Pour Yazdi, M.; Billard, A.; François, M.; Sanchette, F. Ni-free superelastic binary Ti–Nb coatings obtained by DC magnetron co-sputtering. *Surf. Coat. Technol.* **2015**, *275*, 283–288. [CrossRef]

15. Achache, S.; Alhussein, A.; Guelorget, B.; Salut, R.; François, M.; Sanchette, F. Effect of oxygen addition on microstructure and mechanical properties of quaternary TNTZ superelastic thin films obtained by magnetron sputtering. *Mater. Chem. Phys.* **2018**, *217*, 262–269. [CrossRef]

16. Achache, S.; Alhussein, A.; Lamri, S.; François, M.; Sanchette, F.; Pulgarin, C.; Kiwi, J.; Rtimi, S. Sputtered Gum metal thin films showing bacterial inactivation and biocompatibility. *Colloids Surf. B Biointerfaces* **2016**, *146*, 687–691. [CrossRef] [PubMed]

17. Kanagesan, M.; Hashim, S.; Tamilselvan, A.; Ismail, I.; Ahsanul, K. Synthesis, characterization, and cytotoxicity of iron oxide nanoparticles. *Adv. Mat. Sci. Eng.* **2013**, *2013*, 710432. [CrossRef]

18. Alhussein, A.; Achache, S.; Deturche, R.; Sanchette, F.; Pulgarin, C.; Kiwi, J.; Rtimi, S. Beneficial effect of Cu on Ti-Nb-Ta-Zr sputtered uniform/adhesive gum films accelerating bacterial inactivation under indoor visible light. *Colloids Surf. B Biointerfaces* **2017**, *152*, 152–158. [CrossRef]

19. Seddiki, O.; Harnagea, C.; Levesque, I.; Mantovani, D.; Rosei, F. Evidence ofantibacterial activity on titanium surfaces through nanotextures. *Appl. Surf. Sci.* **2014**, *308*, 275–284. [CrossRef]

20. Kadlec, R.; Jakubec, M.; Jaglic, Z. A novel flotation technique for the separation of Non-Adherent microorganisms from a substrate. *Lett. Appl. Microbiol.* **2014**, *58*, 604–609. [CrossRef]

21. Cullity, B.D. *Elements of X-Ray Diffraction*, 2nd ed.; Addison-Wesley: Reading, MA, USA, 1978; p. 292.

22. Bonnefond, A.; Gonzalez, E.; Asua, J.M.; Leiza, J.R.; Kiwi, J.; Pulgarin, C.; Rtimi, S. New evidence for hybrid acrylic/TiO2 films inducing bacterial inactivation under low intensity simulated sunlight. *Colloids Surf. B Biointerfaces* **2015**, *135*, 1–7. [CrossRef] [PubMed]

23. Lamari, F.; Chakroun, I.; Rtimi, S. Assessment of the correlation among antibiotic resistance, adherence to abiotic and biotic surfaces, invasion and cytotoxicity of Pseudomonas aeruginosa isolated from diseased gilthead sea bream. *Colloids Surf. B Biointerfaces* **2017**, *158*, 229–236. [CrossRef] [PubMed]

24. Rtimi, S.; Kiwi, J. Recent advances on sputtered films with Cu in ppm concentrations leading to an acceleration of the bacterial inactivation. *Catal. Today* **2020**, *340*, 347–362. [CrossRef]

25. Rtimi, S. Indoor light enhanced photocatalytic Ultra-Thin films on flexible Non-Heat resistant substrates reducing bacterial infection risks. *Catalysts* **2017**, *7*, 57. [CrossRef]

26. Rtimi, S.; Sanjines, R.; Pulgarin, C.; Kiwi, J. Microstructure of Cu–Ag uniform nanoparticulate films on polyurethane 3D catheters: Surface properties. *ACS Appl. Mater. Interfaces* **2015**, *8*, 56–63. [CrossRef]

27. Graf, A.; Finkel, J.; Chauvet, A.A.P.; Rtimi, S. Deciphering the mechanisms of bacterial inactivation on HiPIMS Sputtered CuxO-FeOx-PET Surfaces: From light absorption to catalytic bacterial death. *ACS Appl. Mater. Interfaces* **2019**, *11*, 45319–45329. [CrossRef]

28. Rtimi, S.; Dionysiou, D.D.; Pillai, S.C.; Kiwi, J. Advances in bacterial inactivation by Ag, Cu, Cu-Ag coated surfaces & medical devices. *Appl. Catal. B Environ.* **2019**, *240*, 291–318.

29. Zeghioud, H.; Assadi, A.A.; Khellaf, N.; Djelal, H.; Amrane, A.; Rtimi, S. Photocatalytic performance of CuxO/TiO2 deposited by HiPIMS on polyester under visible light LEDs: Oxidants, ions effect, and reactive oxygen species investigation. *Materials* **2019**, *12*, 412. [CrossRef]

 *coatings*

Perspective

Nanoengineered Antibacterial Coatings and Materials: A Perspective

Krasimir Vasilev

School of Engineering, The University of South Australia, Mawson Lakes SA 5095, Australia;
krasimir.vasilev@unisa.edu.au

Received: 5 September 2019; Accepted: 9 October 2019; Published: 11 October 2019

Abstract: This feature article begins by outlining the problem of infection and its implication on healthcare. The initial introductory section is followed by a description of the four distinct classes of antibacterial coatings and materials, i.e., bacteria repealing, contact killing, releasing and responsive, that were developed over the years by our team and others. Specific examples of each individual class of antibacterial materials and a discussion on the pros and cons of each strategy are provided. The article contains a dedicated section focused on silver nanoparticle based coatings and materials, which have attracted tremendous interest from the scientific and medical communities. The article concludes with the author's view regarding the future of the field.

Keywords: infection; antibacterial; coatings; silver; nanomaterials; plasma deposition

1. Introduction

Infections have accompanied humans for millennia. Today, in the 21st century, despite being in the era of antibiotics, which began with the discovery of penicillin by Alexander Fleming in 1928, infections are still an important problem to human wellbeing. Antibiotics have saved countless lives and are considered by many as the greatest medical discovery of the 20th century. However, resistance emerged not long after the mass production and use of antibiotics had begun. Today, resistance is a significant issue since many bacterial species are no longer susceptible to commercially available antibiotics [1]. This requires the development of new compounds. However, the process is expensive, time consuming and associated with significant regulatory burdens. For these reasons, almost no new antibiotic compounds have been provided to the market in the last two decades.

Infections are a particularly difficult issue when associated with implantable medical devices [2]. For example, up to 2% of orthopaedic devices, 5% of trauma fixation devices and more that 20% of devices used for endoprosthetic reconstruction of large bone defects will become infected [3]. While infection rates of 1%–2% in the case of orthopaedic devices may not look too dramatic, infections are the reason for nearly 50% of all revisions that are required for this type of device [4]. Catheter related bacteraemia (CRB) is often defined as "a problem of epidemic proportion in the dialysis population" [5] The incidence of catheter related bacteraemia is in the range of 4–6 per 1000 patient-days in Australia, the US and other developed countries. A study conducted in Australia, and published in 2009, reported that 1163 patients out of 21,935 patients (5.1%) who started dialysis therapy died of infections [6]. Among critically ill patients developing catheter-related bacteraemia, mortality rates are as high as 35%, while survivors incurred increased hospital costs of at least $40,000 and increased time spent in the ICU of 8 to 20 days [7].

The majority of medical device associated infections are caused by bacteria, although fungal and mixed infections have also gained recognition in recent years [3,4,8]. In a typical scenario, a medical device becomes contaminated by bacteria before or during the implantation procedure. In a proper clinical setting, this will be a relatively small number of bacteria. Once attached to the device surface, bacteria begins to proliferate and form colonies, express extra cellular products and a biofilm. With time, the biofilm matures and provides an effective protection for the bacteria from antibiotics and the patient's immune system. Bacteria in biofilm also undergo phenotypical changes, which makes them different to their planktonic counterparts and helps them to develop antibiotic resistance [1]. It is broadly acknowledged that bacteria in biofilm can resist 1000 times higher doses of antibiotics than planktonic bacteria. This presents significant challenges for the treatment of infected implantable medical devices since traditional antibiotic therapies can be ineffective [8–10]. In such situations, the only clinical option may be the removal and replacement of the device, which is a costly procedure that causes patient suffering and carries an even higher probability of infection. A further challenge is the population demographics receiving implantable medical devices, which in many cases is elderly people with reduced immune system strength and patients with supressed immune systems due to receiving chemotherapy (oncology patients) or treatments for other diseases.

2. Antibacterial Coatings

Since medical device associated infections often begin with the attachment of a few individual planktonic bacteria to the surface of the device, a concept that has gained tremendous attention from researchers, industry and healthcare professionals is the prevention of the initial bacterial attachment by applying antibacterial coatings. The breadth of research activities in this area is enormous and has been reviewed by us and many other authors [2,11–13]. In this invited feature article, the aim of the author is to summarise and place into perspective the research from our group on nanoengineered antibacterial surfaces and materials that was published over the last decade and is currently under development. The author is not rejecting or undermining the excellent work of many other researchers in the field, he is just adhering to the purpose of this feature paper.

Over the last decade, our team has developed four distinct classes of antibacterial materials and coatings. These classes of materials are schematically depicted in Figure 1. The first class (Figure 1a) of antibacterial coatings are those that bacteria simply do not like to attach to. This type of coating is typically based on hydrophilic polymers, such as polyethylene glycol (PEG), oxazolines, nitroxide radicals or chlorinated plasma polymers [14–18]. A specific section in this article will be attributed to oxazolines. The second class of coatings or modified material surfaces are those which are capable of killing bacteria upon contact (Figure 1b). An example of this coating is surface grafted quaternary ammonium compounds (QAC) [19,20]. Using a carefully designed number density gradient of QAC, our team was able to demonstrate that a surface concentration of RNH_4^+ bonded nitrogen of 4.18% and surface potential of +120.4 mV is required to achieve the efficient killing of *Escherichia coli* [19]. While the strategies presented in Figure 1a,b are great approaches to stopping bacteria colonising on the surface of the device, they do not eliminate bacteria which may have infiltrated the site of implantation. This places an open wound at risk from opportunistic pathogens. To neutralise such pathogens, coatings or materials which release antibacterial agents were developed (Figure 1c). Many antibacterial compounds can be released by this strategy, including conventional antibiotics [21,22], nitric oxide [23–25], antibacterial polymers and peptides [26–28]. A special section of this article will be devoted to silver nanoparticles with an explanation of why such coatings are classified as releasing. The last category comprises of coatings and materials which release antibacterial on demand (Figure 1d), or only when photogenic bacteria have contaminated the device or its vicinity [12,29,30]. These 'intelligent' materials have many advantages compared to the other three classes of antibacterial coatings and will be discussed in a separate section.

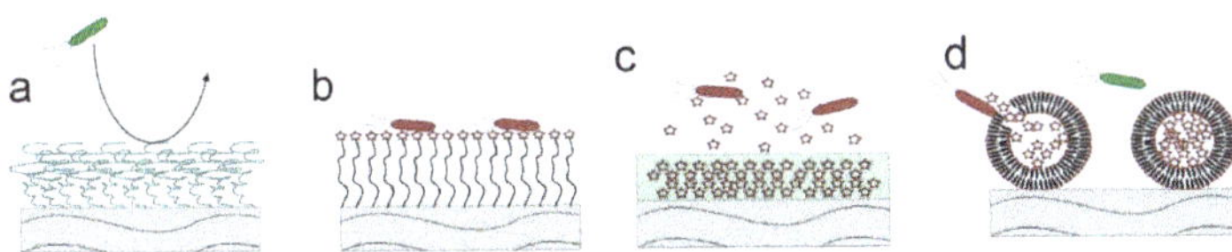

Figure 1. Classes of antibacterial coatings: (**a**) repelling bacterial attachment or biofilm development; (**b**) contact killing; (**c**) releasing of antibacterial agents; and (**d**) stimuli responsive release in the presence of bacteria.

Plasma Deposition

Many of our antibacterial coatings are facilitated by a method called plasma deposition [11,31]. Plasma coatings are deposited from the gaseous form of a selected precursor in which molecules are electrically excited to a plasma state. The nature of the technique offers unique versatility to deposit coatings of a wide range of functional groups, such as amine [32], carboxyl [33], hydroxyl [34], epoxy [35], oxazoline [36], chlorine [17], fluorine [37], siloxane [38], including from precursors that are not polymerizable by conventional means, such as ethanol [39]. The method is fast (seconds to minutes), can be completed in a single step and does not require the use of solvents as in the case of wet methods for surface modification [40,41]. Nanoscale coatings prepared by plasma deposition adhere well to almost any type of substrate material without the need for substrate pre-modification. We have also demonstrated that after a few nanometres of coating are deposited, the film growth becomes substrate independent [42–44]. This is a significant advantage since medical devices are made from all four classes of materials, i.e., polymers, metals, ceramics and composites. Using plasma deposition allows all of these materials to be coated by using the same process and without the need for process optimisation to suit a particular substrate material. In comparison, wet techniques for the preparation of thin films, such as Layer-by-Layer (LbL) or Self Assembled Monolayers (SAMs), are limited to the specific type of substrate material they require [45]. Moreover, plasma polymers can be deposited on substrates of complex shapes, such as porous materials or those having complex nanotopography, as well as micro and nanoparticles [22,24,37,46–50].

In the following, this article will provide a perspective on specific examples of the different classes of antibacterial surfaces depicted in Figure 1.

3. Oxazoline Based Coatings That Inhibit Biofilm Growth

A recently developed class of plasma polymers are those deposited from oxazoline precursors, such as 2-methyl-2-oxazoline and 2-ethyl-2-oxazoline [15,36,51–54]. Regardless of the precursor used, when appropriate conditions for deposition are used, the process results in coatings with properties that are very useful for biomedical applications. Uniquely, the plasma deposition process allows for the retention of intact oxazoline rings on the surface of the coatings, which would normally be lost if polymerisation was carried out by conventional ring opening. These rings are very reactive to carboxyl acid groups in a one-step, click type reaction without the need for any catalysts or intermediates. This special property allows for covalent binding of biomolecules, ligands and nanoparticles containing carboxyl acid groups, which provide platforms for developing diagnostic technologies and tools for interrogating biological phenomena occurring at the material interphase [55–58]. Oxazoline derived plasma polymer coatings were also demonstrated to efficiently modulate the immune responses exhibited by the reduction in secretion of pro-inflammatory cytokines and the composition of protein corona forming on the biomaterial surface [15,56,59]. In the context of this article, the most interesting property of oxazoline based plasma deposited coatings is the inhibition of biofilm formation [14,15,53,60]. An example is shown in Figure 2, which depicts four cell culture wells after incubation in a culture of *S. epidermidis* for 24 h. A surface coated with an amine plasma polymer was used as a positive control. The images clearly show that bacteria were able to form a significant biofilm on the positive control surface (top right). However, when coatings based on oxazoline precursors were applied, the formation of biofilm was completely inhibited (bottom). Microscopy images confirmed these observations at the macroscopic level (shown in the right-hand side of the image). Our team is still interrogating the exact mechanisms by which coating deposited from oxazoline precursors inhibit bacterial growth. We expect to report on this phenomenon in the near future. In the meantime, this type of coating presents an appealing option for the modification of medical device surfaces, such as artificial heart valves, pacemakers and catheters. The fact that such low biofouling coatings can be prepared under atmospheric conditions using a plasma jet adds another dimension of interest [60].

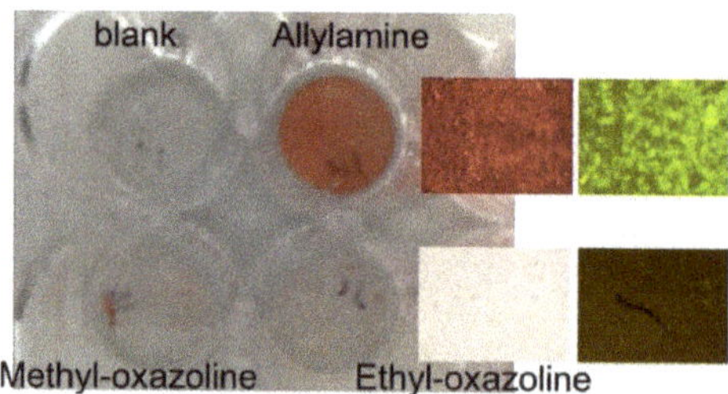

Figure 2. Growth of *S. epidermidis* for 24 h on glass coverslips coated with allylamine, 2-methyl-2-oxazoline and 2-ethyl-2-oxazoline. While biofilm readily forms on allylamine coated surfaces, a complete inhibition of biofilm growth is seen when the same type of substrate is coated with either 2-methyl-2-oxazoline or 2-ethyl-2-oxazoline. The microscopy images confirm the macroscopic observations.

4. Contact Killing Surfaces

Surfaces which are capable of killing bacteria upon contact have generated significant scientific attention. This interest was further stimulated by a report from Ivanova and co-workers, who reported that surface nanopillars resembling dragon fly surface architecture can efficiently kill medically relevant pathogens with great efficiency [61]. We have further added to the field by evaluating the effect of the nanostructure's outermost surface chemistry on antibacterial properties [62]. By applying a thin plasma polymer over layer, we were able to retain the original surface topography and to just fine tune the outermost surface chemistry. We found that hydrophilic plasma polymer coatings enhanced antimicrobial activity, while hydrophobic coatings reduced it [62]. Currently, our team is working with a commercial partner to transfer similar surface nanotopography on orthopaedic implants and bring it to market. As discussed above, QACs can be grafted to plasma polymer coatings to create contact killing surfaces [19,20]. Other potential alternatives for grafting on surfaces are antibacterial polymers and peptides, which have also attracted much attention in the last decade [26–28].

5. Releasing Surfaces

Surfaces and materials which release antibacterial compounds are the most commonly used ones due to their simplicity and efficacy to eliminate pathogens both on the surface of the device and in its vicinity.

5.1. Silver Containing Coatings and Materials

After being almost forgotten in the beginning of the era of antibiotics, silver came back to medicine in the second half of last century due to its antibacterial potency [63]. Materials incorporating silver are classified as releasing because of the way they work. Whether in the form of nanoparticles or continuous metallic layers, silver oxidizes when placed in a physiological medium. The oxide is then dissolved leading to the release of silver ions (Figure 3). These ions are the actual species which kill bacteria by a multifaceted mechanism of action, such as binding to the cell membrane causing lysis, blocking replication by binding to DNA, binding to enzymes and proteins and, in this way, interfering with bacterial metabolism. Silver is capable of killing both Gram-positive and Gram-negative pathogens. Due to such a multifaceted mechanism of action, it is also more difficult for bacteria to develop a resistance to silver compared to conventional antibiotics.

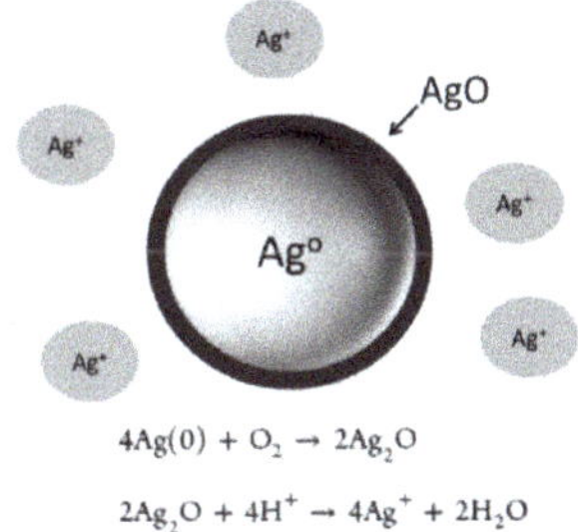

Figure 3. Mechanism for the dissolution of silver nanoparticles in an aqueous environment. The dissolved silver ions are the species killing bacteria.

Our team has done a significant body of work on antibacterial coatings containing silver nanoparticles [64–84]. In a paper published in Nano Letters in 2010, we were one of the first to report that mammalian cells can have greater tolerance to silver than bacteria [85]. These findings opened a therapeutic window for the safe application of silver on medical devices. To do that, we designed a platform which consisted of a 100 nm thick plasma polymer film rich in an amine group (Figure 4). This film was then loaded with silver ions by immersion in silver nitrate. The silver ions were then reduced to silver nanoparticles by immersion in sodium borohydride, which is a commonly used reduction agent. To control the rate of oxidation and dissolution of the silver nanoparticles and the subsequent release of silver ions, we applied a 'barrier' plasma polymer layer on top, which had precise thicknesses of 6, 12 and 18 nm. The thickness of the over layer was selected in a way to allow for sufficient silver ion release, and preserve the complete inhibition of bacterial attachment and biofilm formation. While there was some reduction in attachment of osteoblastic cells on the 'as prepared' silver nanoparticle loaded coating, we found that when the rate of silver ion release was reduced (even by the thinnest over layer) the cells grew and proliferated as if on a control without silver nanoparticles. This result pointed to the opportunity to design coatings for medical devices which are completely antibacterial but allow for normal tissue growth and integration. We also compared the efficacy of these coatings to commercial silver dressings (V.A.C.® GRANUFOAM™, Kinetic Concepts Inc., Texas, SA, US) containing a three micron thick silver layer and determined that the same antibacterial efficacy can be achieved with only a fraction of the amount of silver if nanoparticles are used [80].

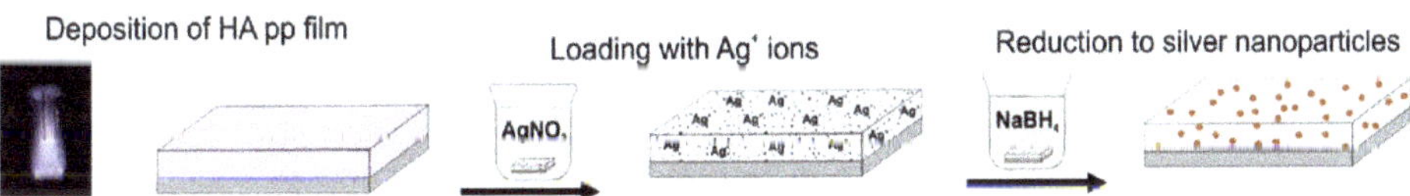

Figure 4. Schematic of the preparation of silver nanoparticles loaded films via first loading silver ions into an amine rich plasma polymer film deposited from the vapour of heptylamine (HA pp) followed by a reduction to silver nanoparticles.

We also prepared silver nanoparticles empowered antibacterial coatings and materials by first synthesizing the nanoparticles in solution and then attaching them to functional plasma polymer coatings (Figure 5) or incorporating them in hybrid materials. The synthetic protocols which we developed involve a range of reduction and capping agents (lipids, 2-mercaptosuccinic acid, citrate, polyvinyl-sulphonate, etc.), including a procedure that benefited from the properties of natural products, such as cacao extract [64,71,73–78,83,86]. These nanoparticles can be immobilised electrostatically or covalently to plasma polymer films containing appropriate surface functionalities (Figure 5) [70,71,75,79,87]. This method allows for easy control of the number of silver nanoparticles immobilised on the surface and can lead to coatings providing full protection against bacterial colonisation, favourable integration of mammalian cells and tissue, and the absence of adverse immune responses. Another way to use these silver nanoparticles is to incorporate them in the material itself, as we did with nanocapsules, nanoparticles and microneedle patches [68,69,84]. Recently, we discovered that ultra-small 1–2 nm silver (and gold) nanoparticles synthesised in an aqueous medium have a very high potency against a range of clinically relevant pathogens, but also reduce expression of pro-inflammatory cytokines from immune cells and help wounds to close [88–90]. We also combined silver nanoparticles and graphene oxide nanostructures and demonstrated a synergic bactericidal effect against Gram-positive and Gram-negative bacteria [81].

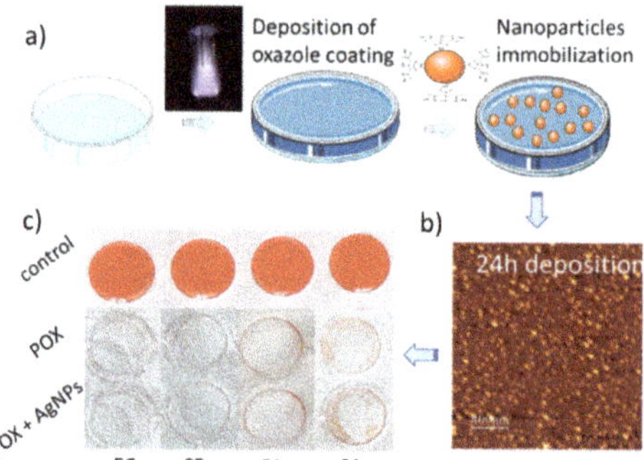

Figure 5. Preparation of silver nanoparticles decorated surfaces and their capacity to inhibit bacterial growth. (**a**) Substrates were first modified by a 20 nm thin layer of oxazoline based plasma polymer (POX). Silver nanoparticles (AgNPs) were synthesised in solution and covalently immobilized by immersion of the substrate for 24 h. (**b**) An AFM (Atomic Force Microscopy) image visualising the silver nanoparticles immobilised to the substrate. (**c**) The POX and POX + AgNPs modified surfaces completely inhibit the growth of *E. coli* (EC), *S. epidermidis* (SE), *P. aeruginosa* (PA) and *S. aureus* (SA), while bacteria easily form biofilm on the control (red colour is staining with Safranin-O). Note that the POX coating is intrinsically non-biofouling (discussed above and shown in Figure 2), while the AgNPs release silver ions to eliminate bacteria in the vicinity of the substrate.

5.2. Release of Conventional Antibiotics, Nitric Oxide and Antibacterial Polymers and Peptides

Antibacterial agents can provide greater efficacy and lower systemic toxicity if released locally. We have developed coatings where levofloxacin was incorporated between two plasma polymer layers. By tuning the thickness of the outermost plasma polymer film, an efficient control of the rate of release could be achieved. Porous materials, such as nanoporous alumina and titania, could be used as reservoirs for loading antibacterials and biologicals [91–93]. Controlled release can be achieved by applying a plasma polymer on top of the structure to reduce the pore openings and, in this way, regulate the dissolution and release of the loaded drugs. Similar strategies could be applied for the release of antibacterial polymers and peptides, which attracted significant attention in the last decade [26–28]. Another interesting compound to be released from coatings or particulates is nitric oxide [23–25]. The molecule has the capacity to inhibit biofilm formation, however, it does not kill the bacteria and the antibacterial effect persists only until NO is released.

6. Responsive Coatings

Intelligent coatings and materials capable of selective release of antibacterial compounds only in the presence of bacteria are of substantial scientific and technological interest. Such materials are a major advance compared to the traditional release of antimicrobial agents, since toxicity to tissue is completely avoided if pathogens have not infiltrated the site of the device. Stimuli for responsive release could be pH, temperature, reactive oxygen species (ROS) or enzymes, and toxins expressed by the bacteria [12] (Figure 6). We have created enzyme responsive nanoparticles and nanocapsules [29,30]. For example, in Reference [29], we synthesized hyaluronic acid nanocapsules containing polyhexanide. Pathogenic bacteria such as *S. aureus* express a number of toxins including an enzyme called hyaluronidase. We demonstrated that the nanocapsules would release their load only in the presence of this specific enzyme. The nanocapsules could also be loaded with fluorescent dyes. Thus, the nanomaterial not only kills the bacteria but also signals for their presence, in this manner informing the medical professional that attention is required.

Figure 6. The core-shell architecture of a responsive (nano)material and the possible constituents of the core and shell.

7. Future Outlook and Challenges

The field of antibacterial coating technology has entered an exciting phase with new technologies continually emerging. The field will only continue to grow in the future since there is increasing recognition of the problem of infection by the medical community, leading to a strong technology pull from the biomedical device industry. As mentioned earlier in this article, and exemplified by orthopaedic devices, infections have become the main reason for device revision. After decades of development, devices such as knee and hip implants are quite sophisticated in terms of design and material selection, causing very few problems. However, infection rates have remained unchanged for decades. There is recognition by the community that a solution for this problem is urgently required. A device surface that is engineered to inhibit attachment of bacteria and biofilm formation could be an ideal solution. However, when such a surface is designed, specific care has to be taken not only regarding safety, but also the regulatory pathways that the new device has to go through. This is relevant for a range of medical devices, such as catheters, wound dressings, pacemakers, stents, tooth implants and stents, just to name a few. Antibacterial surfaces are needed not only in healthcare, but also in many other fields, good examples being the food and marine industries.

The author believes that intelligent materials and coatings that release antibacterial agents only when required and signal for the presence of pathogens will become a main area of development. Such materials would be applicable to many medical devices but particularly for items that are replaced on a relatively short-term basis, as for example wound dressings and catheters. The challenge is to demonstrate that in the absence of bacteria the material is capable of preventing release for substantially longer periods of time (six months or longer for dialysis catheters), but still responds quickly if stimuli are present. This leads to the challenge of selecting the specific stimuli. While temperature and pH are commonly used, it is difficult to design materials that are capable of responding sharply to very small changes in these two parameters in relevant ranges. Responsiveness to selective enzymes, such as lipases, which are expressed by pathogenic bacteria is an attractive option. However, since lipases are also expressed by some human cells, the material should be able to tolerate the natural variation in the level of these enzymes in the body when bacterial are not present. The same is valid for reactive oxygen species, which are released by immune cells to fight pathogens. If this fine balance is achieved, the immune response can be used to enhance the efficacy of the material by accelerating the release rate. There is a huge potential and market for such intelligent materials and our group is directing significant effort and resources to this area.

Funding: This research was funded by ARC (DP15104212, DP DP180101254), NHMRC (Fellowship APP1122825 and Project grant APP1032738), and the Alexander von Humboldt Foundation (Fellowship for Experienced Researchers).

Conflicts of Interest: The authors declare no conflict of interest.

References

1. Campoccia, D.; Montanaro, L.; Arciola, C.R. The significance of infection related to orthopedic devices and issues of antibiotic resistance. *Biomaterials* **2006**, *27*, 2331–2339. [CrossRef] [PubMed]
2. Vasilev, K.; Cook, J.; Griesser, H.J. Antibacterial surfaces for biomedical devices. *Expert Rev. Med. Devices* **2009**, *6*, 553–567. [CrossRef] [PubMed]
3. Bozic, K.; Kurtz, S.M.; Lau, E.; Ong, K.; Vail, T.P.; Rubash, H.E.; Berry, D.J. The epidemiology of revision total knee arthroplasty in the United States. *J. Arthroplast.* **2009**, *24*, e49. [CrossRef]
4. Kurtz, S.M.; Lau, E.; Watson, H.; Schmier, J.K.; Parvizi, J. Economic burden of periprosthetic joint infection in the United States. *J. Arthroplast.* **2012**, *27*, 61–65. [CrossRef] [PubMed]
5. Bleyer, A.J. Use of antimicrobial catheter lock solutions to prevent catheter-related bacteremia. *Clin. J. Am. Soc. Nephrol.* **2007**, *2*, 1073–1078. [CrossRef] [PubMed]
6. Johnson, D.W.; Dent, H.; Hawley, C.M.; McDonald, S.P.; Rosman, J.B.; Brown, F.G.; Bannister, K.M.; Wiggins, K.J. Associations of dialysis modality and infectious mortality in incident dialysis patients in Australia and New Zealand. *Am. J. Kidney Dis.* **2009**, *53*, 290–297. [CrossRef] [PubMed]
7. Dimick, J.B.; Pelz, R.K.; Consunji, R.; Swoboda, S.M.; Hendrix, C.W.; Lipsett, P.A. Increased resource use associated with catheter-related bloodstream infection in the surgical intensive care unit. *Arch. Surg.* **2001**, *136*, 229–234. [CrossRef] [PubMed]
8. Darouiche, R.O. Current concepts—Treatment of infections associated with surgical implants. *N. Engl. J. Med.* **2004**, *350*, 1422–1429. [CrossRef] [PubMed]
9. Kuehl, R.; Brunetto, P.S.; Woischnig, A.K.; Varisco, M.; Rajacic, Z.; Vosbeck, J.; Terracciano, L.; Fromm, K.M.; Khanna, N. Preventing implant-associated infections by silver coating. *Antimicrob. Agents Chemother.* **2016**, *60*, 2467–2475. [CrossRef] [PubMed]
10. Zimmerli, W.; Trampuz, A.; Ochsner, P.E. Current concepts: Prosthetic-joint infections. *N. Engl. J. Med.* **2004**, *351*, 1645–1654. [CrossRef]
11. Vasilev, K.; Griesser, S.S.; Griesser, H.J. Antibacterial surfaces and coatings produced by plasma techniques. *Plasma Process. Polym.* **2011**, *8*, 1010–1023. [CrossRef]
12. Cavallaro, A.; Taheri, S.; Vasilev, K. Responsive and "smart" antibacterial surfaces: Common approaches and new developments (Review). *Biointerphases* **2014**, *9*, 029005. [CrossRef] [PubMed]
13. Chernousova, S.; Epple, M. Silver as antibacterial agent: Ion, nanoparticle, and metal. *Angew. Chem. Int. Ed.* **2013**, *52*, 1636–1653. [CrossRef] [PubMed]
14. Cavallaro, A.A.; MacGregor-Ramiasa, M.N.; Vasilev, K. Antibiofouling properties of plasma-deposited Oxazoline-based thin films. *ACS Appl. Mater. Interfaces* **2016**, *8*, 6354–6362. [CrossRef] [PubMed]
15. Ramiasa, M.N.; Cavallaro, A.A.; Mierczynska, A.; Christo, S.N.; Gleadle, J.M.; Hayball, J.D.; Vasilev, K. Plasma polymerised polyoxazoline thin films for biomedical applications. *Chem. Commun.* **2015**, *51*, 4279–4282. [CrossRef] [PubMed]
16. Michl, T.D.; Coad, B.R.; Hüsler, A.; Valentin, J.D.P.; Vasilev, K.; Griesser, H.J. Effects of precursor and deposition conditions on prevention of bacterial biofilm growth on Chlorinated plasma polymers. *Plasma Process. Polym.* **2016**, *13*, 654–662. [CrossRef]
17. Michl, T.D.; Coad, B.R.; Doran, M.; Hüsler, A.; Valentin, J.D.P.; Vasilev, K.; Griesser, H.J. Plasma polymerization of 1,1,1-trichloroethane yields a coating with robust antibacterial surface properties. *RSC Adv.* **2014**, *4*, 27604–27606. [CrossRef]
18. Michl, T.D.; Barz, J.; Giles, C.; Haupt, M.; Henze, J.H.; Mayer, J.; Futrega, K.; Doran, M.R.; Oehr, C.; Vasilev, K.; et al. Plasma polymerization of TEMPO yields coatings containing stable Nitroxide radicals for controlling interactions with Prokaryotic and Eukaryotic Cells. *ACS Appl. Nano Mater.* **2018**, *1*, 6587–6595. [CrossRef]

19. Cavallaro, A.; Mierczynska, A.; Barton, M.; Majewski, P.; Vasilev, K. Influence of immobilized quaternary ammonium group surface density on antimicrobial efficacy and cytotoxicity. *Biofouling* **2016**, *32*, 13–24. [CrossRef]

20. Cavallaro, A.; Majewski, P.; Barton, M.; Vasilev, K. Substrate independent approach for immobilisation of quaternary ammonium compounds to surfaces to reduce bio-burden. *Mater. Sci. Forum* **2014**, *783*, 1389–1395. [CrossRef]

21. Vasilev, K.; Poulter, N.; Martinek, P.; Griesser, H.J. Controlled release of levofloxacin sandwiched between two plasma polymerized layers on a solid carrier. *ACS Appl. Mater. Interfaces* **2011**, *3*, 4831–4836. [CrossRef] [PubMed]

22. Cavallaro, A.; Vasilev, K. Controlled and sustained release of pharmaceuticals via single step solvent-free encapsulation. *Chem. Commun.* **2015**, *51*, 1838–1841. [CrossRef] [PubMed]

23. Michl, T.D.; Coad, B.R.; Doran, M.; Osiecki, M.; Kafshgari, M.H.; Voelcker, N.H.; Hüsler, A.; Vasilev, K.; Griesser, H.J. Nitric oxide releasing plasma polymer coating with bacteriostatic properties and no cytotoxic side effects. *Chem. Commun.* **2015**, *51*, 7058–7060. [CrossRef] [PubMed]

24. Kafshgari, M.H.; Cavallaro, A.; Delalat, B.; Harding, F.J.; McInnes, S.J.P.; Mäkilä, E.; Salonen, J.; Vasilev, K.; Voelcker, N.H. Nitric oxide-releasing porous silicon nanoparticles. *Nanoscale Res. Lett.* **2014**, *9*, 333. [CrossRef] [PubMed]

25. Hasanzadeh Kafshgari, M.; Delalat, B.; Harding, F.J.; Cavallaro, A.; Mäkilä, E.; Salonen, J.; Vasilev, K.; Voelcker, N.H. Antibacterial properties of nitric oxide-releasing porous silicon nanoparticles. *J. Mater. Chem. B* **2016**, *4*, 2051–2058. [CrossRef]

26. Michl, T.D.; Locock, K.E.S.; Stevens, N.E.; Hayball, J.D.; Vasilev, K.; Postma, A.; Qu, Y.; Traven, A.; Haeussler, M.; Meagher, L.; et al. RAFT-derived antimicrobial polymethacrylates: Elucidating the impact of end-groups on activity and cytotoxicity. *Polym. Chem.* **2014**, *5*, 5813–5822. [CrossRef]

27. Locock, K.E.S.; Michl, T.D.; Valentin, J.D.P.; Vasilev, K.; Hayball, J.D.; Qu, Y.; Traven, A.; Griesser, H.J.; Meagher, L.; Haeussler, M. Guanylated polymethacrylates: A class of potent antimicrobial polymers with low hemolytic activity. *Biomacromolecules* **2013**, *14*, 4021–4031. [CrossRef] [PubMed]

28. Locock, K.E.S.; Michl, T.D.; Stevens, N.; Hayball, J.D.; Vasilev, K.; Postma, A.; Griesser, H.J.; Meagher, L.; Haeussler, M. Antimicrobial polymethacrylates synthesized as mimics of tryptophan-rich cationic peptides. *ACS Macro Lett.* **2014**, *3*, 319–323. [CrossRef]

29. Baier, G.; Cavallaro, A.; Vasilev, K.; Mailänder, V.; Musyanovych, A.; Landfester, K. Enzyme responsive hyaluronic acid nanocapsules containing polyhexanide and their exposure to bacteria to prevent infection. *Biomacromolecules* **2013**, *14*, 1103–1112. [CrossRef]

30. Baier, G.; Cavallaro, A.; Friedemann, K.; Müller, B.; Glasser, G.; Vasilev, K.; Landfester, K. Enzymatic degradation of poly(L-lactide) nanoparticles followed by the release of octenidine and their bactericidal effects. *Nanomed. Nanotechnol. Biol. Med.* **2014**, *10*, 131–139. [CrossRef]

31. Vasilev, K. Nanoengineered plasma polymer films for biomaterial applications. *Plasma Chem. Plasma Process.* **2014**, *34*, 545–558. [CrossRef]

32. Liu, X.; Shi, S.; Feng, Q.; Bachhuka, A.; He, W.; Huang, Q.; Zhang, R.; Yang, X.; Vasilev, K. Surface chemical gradient affects the differentiation of human adipose-derived stem cells via ERK1/2 signaling pathway. *ACS Appl. Mater. Interfaces* **2015**, *7*, 18473–18482. [CrossRef] [PubMed]

33. Mierczynska, A.; Michelmore, A.; Tripathi, A.; Goreham, R.V.; Sedev, R.; Vasilev, K. PH-tunable gradients of wettability and surface potential. *Soft Matter* **2012**, *8*, 8399–8404. [CrossRef]

34. Hazrati, H.D.; Whittle, J.D.; Vasilev, K. A mechanistic study of the plasma polymerization of ethanol. *Plasma Process. Polym.* **2014**, *11*, 149–157. [CrossRef]

35. Thierry, B.; Jasieniak, M.; De Smet, L.C.P.M.; Vasilev, K.; Griesser, H.J. Reactive epoxy-functionalized thin films by a pulsed plasma polymerization process. *Langmuir* **2008**, *24*, 10187–10195. [CrossRef]

36. MacGregor, M.N.; Michelmore, A.; Safizadeh Shirazi, H.; Whittle, J.; Vasilev, K. Secrets of plasma deposited polyoxazoline functionality lie in the plasma phase. *Chem. Mater.* **2017**, *29*, 8047–8051. [CrossRef]

37. Michl, T.D.; Coad, B.R.; Hüsler, A.; Vasilev, K.; Griesser, H.J. Laboratory scale systems for the plasma treatment and coating of particles. *Plasma Process. Polym.* **2015**, *12*, 305–313. [CrossRef]

38. Michelmore, A.; Bryant, P.M.; Steele, D.A.; Vasilev, K.; Bradley, J.W.; Short, R.D. Role of positive ions in determining the deposition rate and film chemistry of continuous wave hexamethyl disiloxane plasmas. *Langmuir* **2011**, *27*, 11943–11950. [CrossRef]

39. Coad, B.R.; Vasilev, K.; Diener, K.R.; Hayball, J.D.; Short, R.D.; Griesser, H.J. Immobilized streptavidin gradients as bioconjugation platforms. *Langmuir* **2012**, *28*, 2710–2717. [CrossRef]

40. Goreham, R.V.; Mierczynska, A.; Pierce, M.; Short, R.D.; Taheri, S.; Bachhuka, A.; Cavallaro, A.; Smith, L.E.; Vasilev, K. A substrate independent approach for generation of surface gradients. *Thin Solid Films.* **2013**, *528*, 106–110. [CrossRef]

41. Coad, B.R.; Scholz, T.; Vasilev, K.; Hayball, J.D.; Short, R.D.; Griesser, H.J. Functionality of proteins bound to plasma polymer surfaces. *ACS Appl. Mater. Interfaces* **2012**, *4*, 2455–2463. [CrossRef] [PubMed]

42. Vasilev, K.; Michelmore, A.; Martinek, P.; Chan, J.; Sah, V.; Griesser, H.J.; Short, R.D. Early stages of growth of plasma polymer coatings deposited from nitrogen- and oxygen-containing monomers. *Plasma Process. Polym.* **2010**, *7*, 824–835. [CrossRef]

43. Vasilev, K.; Michelmore, A.; Griesser, H.J.; Short, R.D. Substrate influence on the initial growth phase of plasma-deposited polymer films. *Chem. Commun.* **2009**, *24*, 3600–3602. [CrossRef] [PubMed]

44. Michelmore, A.; Martinek, P.; Sah, V.; Short, R.D.; Vasilev, K. Surface morphology in the early stages of plasma polymer film growth from amine-containing monomers. *Plasma Process. Polym.* **2011**, *8*, 367–372. [CrossRef]

45. Hernandez-Lopez, J.L.; Bauer, R.E.; Chang, W.S.; Glasser, G.; Grebel-Koehler, D.; Klapper, M.; Kreiter, M.; Leclaire, J.; Majoral, J.P.; Mittler, S.; et al. Functional polymers as nanoscopic building blocks. *Mater. Sci. Eng. C* **2003**, *23*, 267–274. [CrossRef]

46. McInnes, S.J.P.; Michl, T.D.; Delalat, B.; Al-Bataineh, S.A.; Coad, B.R.; Vasilev, K.; Griesser, H.J.; Voelcker, N.H. "Thunderstruck": Plasma-Polymer-Coated Porous Silicon Microparticles As a Controlled Drug Delivery System. *ACS Appl. Mater. Interfaces* **2016**, *8*, 4467–4476. [CrossRef] [PubMed]

47. Wahono, S.K.; Cavallaro, A.; Vasilev, K.; Mierczynska, A. Plasma polymer facilitated magnetic technology for removal of oils from contaminated waters. *Environ. Pollut.* **2018**, *240*, 725–732. [CrossRef]

48. Mierczynska-Vasilev, A.; Boyer, P.; Vasilev, K.; Smith, P.A. A novel technology for the rapid, selective, magnetic removal of pathogenesis-related proteins from wines. *Food Chem.* **2017**, *232*, 508–514. [CrossRef]

49. Ramiasa-MacGregor, M.; Mierczynska, A.; Sedev, R.; Vasilev, K. Tuning and predicting the wetting of nanoengineered material surface. *Nanoscale* **2016**, *8*, 4635–4642. [CrossRef]

50. Christo, S.N.; Bachhuka, A.; Diener, K.R.; Mierczynska, A.; Hayball, J.D.; Vasilev, K. The Role of Surface Nanotopography and Chemistry on Primary Neutrophil and Macrophage Cellular Responses. *Adv. Healthc. Mater.* **2016**, *5*, 956–965. [CrossRef]

51. Macgregor-Ramiasa, M.N.; Cavallaro, A.A.; Vasilev, K. Properties and reactivity of polyoxazoline plasma polymer films. *J. Mater. Chem. B* **2015**, *3*, 6327–6337. [CrossRef]

52. MacGregor, M.; Sinha, U.; Visalakshan, R.M.; Cavallaro, A.; Vasilev, K. Preserving the reactivity of coatings plasma deposited from oxazoline precursors—An in depth study. *Plasma Process. Polym.* **2019**, *16*, 1800130. [CrossRef]

53. Macgregor, M.; Vasilev, K. Perspective on plasma polymers for applied biomaterials nanoengineering and the recent rise of oxazolines. *Materials* **2019**, *12*, 191. [CrossRef] [PubMed]

54. Vasilev, K. Plasma derived oxazoline based coatings for advanced medical technologies. *Galvanotechnik* **2019**, *110*, 170–175.

55. Gonzalez Garcia, L.E.; Macgregor-Ramiasa, M.; Visalakshan, R.M.; Vasilev, K. Protein Interactions with Nanoengineered Polyoxazoline Surfaces Generated via Plasma Deposition. *Langmuir* **2017**, *33*, 7322–7331. [CrossRef] [PubMed]

56. Chen, Z.; Visalakshan, R.M.; Guo, J.; Wei, F.; Zhang, L.; Chen, L.; Lin, Z.; Vasilev, K.; Xiao, Y. Plasma deposited poly-oxazoline nanotextured surfaces dictate osteoimmunomodulation towards ameliorative osteogenesis. *Acta Biomater.* **2019**, *96*, 568–581. [CrossRef] [PubMed]

57. Macgregor-Ramiasa, M.; McNicholas, K.; Ostrikov, K.; Li, J.; Michael, M.; Gleadle, J.M.; Vasilev, K. A platform for selective immuno-capture of cancer cells from urine. *Biosens. Bioelectron.* **2017**, *96*, 373–380. [CrossRef]

58. Visalakshan, R.M.; MacGregor, M.N.; Cavallaro, A.A.; Sasidharan, S.; Bachhuka, A.; Mierczynska-Vasilev, A.M.; Hayball, J.D.; Vasilev, K. Creating Nano-engineered Biomaterials with Well-Defined Surface Descriptors. *ACS Appl. Nano Mater.* **2018**, *1*, 2796–2807. [CrossRef]

59. Visalakshan, R.M.; MacGregor, M.N.; Sasidharan, S.; Ghazaryan, A.; Mierczynska-Vasilev, A.M.; Morsbach, S.; Mailänder, V.; Landfester, K.; Hayball, J.D.; Vasilev, K. Biomaterial Surface Hydrophobicity-Mediated Serum Protein Adsorption and Immune Responses. *ACS Appl. Mater. Interfaces* **2019**. [CrossRef]

60. Al-Bataineh, S.A.; Cavallaro, A.A.; Michelmore, A.; Macgregor, M.N.; Whittle, J.D.; Vasilev, K. Deposition of 2-oxazoline-based plasma polymer coatings using atmospheric pressure helium plasma jet. *Plasma Process. Polym.* **2019**. [CrossRef]

61. Ivanova, E.P.; Hasan, J.; Webb, H.K.; Gervinskas, G.; Juodkazis, S.; Truong, V.K.; Wu, A.H.; Lamb, R.N.; Baulin, V.A.; Watson, G.S.; et al. Bactericidal activity of black silicon. *Nat. Commun.* **2013**, *4*, 2838. [CrossRef]

62. Ostrikov, K.; MacGregor-Ramiasa, M.; Cavallaro, A.; Vasilev, K. Bactericidal effects of plasma-modified surface chemistry of silicon nanograss. *J. Phys. D Appl. Phys.* **2016**, *49*, 304001. [CrossRef]

63. Alexander, J.W. History of the medical use of silver. *Surg. Infect.* **2009**, *10*, 289–292. [CrossRef]

64. Vasilev, K.; Sah, V.R.; Goreham, R.V.; Ndi, C.; Short, R.D.; Griesser, H.J. Antibacterial surfaces by adsorptive binding of polyvinyl-sulphonate-stabilized silver nanoparticles. *Nanotechnology* **2010**, *21*, 215102. [CrossRef] [PubMed]

65. Ploux, L.; Mateescu, M.; Anselme, K.; Vasilev, K. Antibacterial properties of silver-loaded plasma polymer coatings. *J. Nanomater.* **2012**, *2012*, 6. [CrossRef]

66. Lombi, E.; Donner, E.; Taheri, S.; Tavakkoli, E.; Jämting, Å.K.; McClure, S.; Naidu, R.; Miller, B.W.; Scheckel, K.G.; Vasilev, K. Transformation of four silver/silver chloride nanoparticles during anaerobic treatment of wastewater and post-processing of sewage sludge. *Environ. Pollut.* **2013**, *176*, 193–197. [CrossRef] [PubMed]

67. Poulter, N.; Vasilev, K.; Griesser, S.S.; Griesser, H.J. Silver containing biomaterials. In *Biomaterials Associated Infection: Immunological Aspects and Antimicrobial Strategies*; Springer: Berlin/Heidelberg, Germany, 2013; pp. 355–378.

68. Taheri, S.; Baier, G.; Majewski, P.; Barton, M.; Förch, R.; Landfester, K.; Vasilev, K. Synthesis and surface immobilization of antibacterial hybrid silver-poly(L-lactide) nanoparticles. *Nanotechnology* **2014**, *25*, 305102. [CrossRef] [PubMed]

69. Taheri, S.; Baier, G.; Majewski, P.; Barton, M.; Förch, R.; Landfester, K.; Vasilev, K. Synthesis and antibacterial properties of a hybrid of silver-potato starch nanocapsules by miniemulsion/polyaddition polymerization. *J. Mater. Chem. B* **2014**, *2*, 1838–1845. [CrossRef]

70. Taheri, S.; Cavallaro, A.; Barton, M.; Whittle, J.D.; Majewski, P.; Smith, L.E.; Vasilev, K. Antibacterial efficacy and cytotoxicity of silver-nanoparticle–based coatings facilitated by a plasma deposited polymer interlayer. *Plasma Med.* **2014**, *4*, 101–115. [CrossRef]

71. Taheri, S.; Cavallaro, A.; Christo, S.N.; Smith, L.E.; Majewski, P.; Barton, M.; Hayball, J.D.; Vasilev, K. Substrate independent silver nanoparticle based antibacterial coatings. *Biomaterials* **2014**, *35*, 4601–4609. [CrossRef]

72. Brunetti, G.; Donner, E.; Laera, G.; Sekine, R.; Scheckel, K.G.; Khaksar, M.; Vasilev, K.; De Mastro, G.; Lombi, E. Fate of zinc and silver engineered nanoparticles in sewerage networks. *Water Res.* **2015**, *77*, 72–84. [CrossRef] [PubMed]

73. Khaksar, M.; Jolley, D.F.; Sekine, R.; Vasilev, K.; Johannessen, B.; Donner, E.; Lombi, E. In situ chemical transformations of silver nanoparticles along the water-sediment continuum. *Environ. Sci. Technol.* **2015**, *49*, 318–325. [CrossRef] [PubMed]

74. Sekine, R.; Khurana, K.; Vasilev, K.; Lombi, E.; Donner, E. Quantifying the adsorption of ionic silver and functionalized nanoparticles during ecotoxicity testing: Test container effects and recommendations. *Nanotoxicology* **2015**, *9*, 1005–1012. [CrossRef] [PubMed]

75. Taheri, S.; Cavallaro, A.; Christo, S.N.; Majewski, P.; Barton, M.; Hayball, J.D.; Vasilev, K. Antibacterial Plasma Polymer Films Conjugated with Phospholipid Encapsulated Silver Nanoparticles. *ACS Biomater. Sci. Eng.* **2015**, *1*, 1278–1286. [CrossRef]

76. Taheri, S.; Vasilev, K.; Majewski, P. Silver nanoparticles: Synthesis, antimicrobial coatings, and applications for medical devices. *Recent Pat. Mater. Sci.* **2015**, *8*, 166–175. [CrossRef]

77. Alhmoud, H.; Delalat, B.; Ceto, X.; Elnathan, R.; Cavallaro, A.; Vasilev, K.; Voelcker, N.H. Antibacterial properties of silver dendrite decorated silicon nanowires. *RSC Adv.* **2016**, *6*, 65976–65987. [CrossRef]

78. Chowdhury, N.R.; MacGregor-Ramiasa, M.; Zilm, P.; Majewski, P.; Vasilev, K. 'Chocolate' silver nanoparticles: Synthesis, antibacterial activity and cytotoxicity. *J. Colloid Interface Sci.* **2016**, *482*, 151–158. [CrossRef]

79. He, W.; Elkhooly, T.A.; Liu, X.; Cavallaro, A.; Taheri, S.; Vasilev, K.; Feng, Q. Silver nanoparticle based coatings enhance adipogenesis compared to osteogenesis in human mesenchymal stem cells through oxidative stress. *J. Mater. Chem. B* **2016**, *4*, 1466–1479. [CrossRef]

80. Ostrikov, K.; Macgregor-Ramiasa, M.N.; Cavallaro, A.A.; Jacob, M.; Vasilev, K. A comparative assessment of nanoparticulate and metallic silver coated dressings. *Recent Pat. Mater. Sci.* **2016**, *9*, 50–57. [CrossRef]

81. Prasad, K.; Lekshmi, G.S.; Ostrikov, K.; Lussini, V.; Blinco, J.; Mohandas, M.; Vasilev, K.; Bottle, S.; Bazaka, K.; Ostrikov, K. Synergic bactericidal effects of reduced graphene oxide and silver nanoparticles against Gram-positive and Gram-negative bacteria. *Sci. Rep.* **2017**, *7*, 1591. [CrossRef]

82. Schmidt-Braekling, T.; Streitbuerger, A.; Gosheger, G.; Boettner, F.; Nottrott, M.; Ahrens, H.; Dieckmann, R.; Guder, W.; Andreou, D.; Hauschild, G. Silver-coated megaprostheses: Review of the literature. *Eur. J. Orthop. Surg. Traumatol.* **2017**, *27*, 483–489. [CrossRef] [PubMed]

83. Roy Chowdhury, N.; Hopp, I.; Zilm, P.; Murray, P.; Vasilev, K. Silver nanoparticle modified surfaces induce differentiation of mouse kidney-derived stem cells. *RSC Adv.* **2018**, *8*, 20334–20340. [CrossRef]

84. González García, L.E.; MacGregor, M.N.; Visalakshan, R.M.; Ninan, N.; Cavallaro, A.A.; Trinidad, A.D.; Zhao, Y.; Hayball, A.J.D.; Vasilev, K. Self-sterilizing antibacterial silver-loaded microneedles. *Chem. Commun.* **2019**, *55*, 171–174. [CrossRef] [PubMed]

85. Vasilev, K.; Sah, V.; Anselme, K.; Ndi, C.; Mateescu, M.; Dollmann, B.; Martinek, P.; Ys, H.; Ploux, L.; Griesser, H.J. Tunable antibacterial coatings that support mammalian cell growth. *Nano Lett.* **2010**, *10*, 202–207. [CrossRef] [PubMed]

86. Chowdhury, N.R.; Cowin, A.J.; Zilm, P.; Vasilev, K. "Chocolate" gold nanoparticles—One pot synthesis and biocompatibility. *Nanomaterials* **2018**, *8*, 496. [CrossRef]

87. Taheri, S.; Ruiz, J.C.; Michelmore, A.; Macgregor, M.; Förch, R.; Majewski, P.; Vasilev, K. Binding of Nanoparticles to Aminated Plasma Polymer Surfaces is Controlled by Primary Amine Density and Solution pH. *J. Phys. Chem. C* **2018**, *122*, 14986–14995. [CrossRef]

88. Biswas, B.; Warr, L.N.; Hilder, E.F.; Goswami, N.; Rahman, M.M.; Churchman, J.G.; Vasilev, K.; Pan, G.; Naidu, R. Biocompatible functionalisation of nanoclays for improved environmental remediation. *Chem. Soc. Rev.* **2019**, *48*, 3740–3770. [CrossRef]

89. Goswami, N.; Bright, R.; Visalakshan, R.M.; Biswas, B.; Zilm, P.; Vasilev, K. Core-in-cage structure regulated properties of ultra-small gold nanoparticles. *Nanoscale Adv.* **2019**, *1*, 2356–2364. [CrossRef]

90. Ravindran Girija, A.; Balasubramanian, S.; Bright, R.; Cowin, A.J.; Goswami, N.; Vasilev, K. Ultrasmall Gold Nanocluster Based Antibacterial Nanoaggregates for Infectious Wound Healing. *ChemNanoMat* **2019**. [CrossRef]

91. Simovic, S.; Losic, D.; Vasilev, K. Controlled drug release from porous materials by plasma polymer deposition. *Chem. Commun.* **2010**, *46*, 1317–1319. [CrossRef]

92. Simovic, S.; Diener, K.R.; Bachhuka, A.; Kant, K.; Losic, D.; Hayball, J.D.; Brownc, M.P.; Vasilev, K. Controlled release and bioactivity of the monoclonal antibody rituximab from a porous matrix: A potential in situ therapeutic device. *Mater. Lett.* **2014**, *130*, 210–214. [CrossRef]

93. Müller, S.; Cavallaro, A.; Vasilev, K.; Voelcker, N.H.; Schönherr, H. Temperature-Controlled Antimicrobial Release from Poly(diethylene glycol methylether methacrylate)-Functionalized Bottleneck-Structured Porous Silicon for the Inhibition of Bacterial Growth. *Macromol. Chem. Phys.* **2016**, *217*, 2243–2251. [CrossRef]

MDPI

St. Alban-Anlage 66

4052 Basel

Switzerland

Tel. +41 61 683 77 34

Fax +41 61 302 89 18

www.mdpi.com

Coatings Editorial Office

E-mail: coatings@mdpi.com

www.mdpi.com/journal/coatings